Solar Module Packaging

Polymeric Requirements and Selection

Solar Module Packaging

Polymeric Requirements and Selection

Michelle Poliskie

CRC Press is an imprint of the
Taylor & Francis Group, an **informa** business

CRC Press
Taylor & Francis Group
6000 Broken Sound Parkway NW, Suite 300
Boca Raton, FL 33487-2742

First issued in paperback 2017

ISBN 13: 978-1-138-07765-2 (pbk)
ISBN 13: 978-1-4398-5072-5 (hbk)

Visit the Taylor & Francis Web site at
http://www.taylorandfrancis.com

and the CRC Press Web site at
http://www.crcpress.com

Contents

Preface

The core technology of a photovoltaic (PV) company is the PV cell, a semiconductor material responsible for turning light into electricity. Despite the importance of this technology, most PV companies currently do not manufacture their solar cells within the United States. In fact, PV modules, also known as panels, are the larger portion of U.S. PV exports. The module is composed of a series of electrically connected solar cells packaged in glass, polymers, and typically, a metallic frame. Currently, only two companies manufacture both the cell and module in the United States [1]. Due to this anemic manufacturing presence, the U.S. government has passed legislation to promote growth. Specifically, after the U.S. financial market tightened in 2007, the Obama administration and Congress passed the American Recovery and Reinvestment Act of 2009. As part of this legislation, the Department of Energy (DOE) was allowed to award loans, grants, and projects to create economic growth in the renewable energy sector. Recognizing the importance of solar cell technology to PV manufacturing, the DOE revised the Buy American provisions to favor government investment in products from those companies with the largest amount of domestic manufacturing. Despite this new legislation, increasing cell manufacturing in the United States is a daunting task, because the U.S. PV industry has already lost its technological advantage. This technological deficit occurred decades ago when the U.S. semiconductor industry began offshoring its manufacturing capabilities.

The semiconductors used to make solar cells are similar to those used to make integrated circuits. Integrated circuits, commonly referred to as chips, are the core technology for various electronics, such as cell phones, flash drives, and computers. While American manufacturing once dominated semiconductor production, today the highest volume of semiconductors comes from Malaysia, Taiwan, and China. The erosion of American manufacturing has been followed by the depletion of research and development (R&D) investments. Today, the majority of U.S. firms perform their R&D overseas in close proximity to their manufacturing lines [2]. With both the semiconductor innovation and production offshored, the United States is at a distinct disadvantage by designing future economic growth around improvements to PV cell technology. However, a potential competitive advantage does exist if there is innovation in other aspects of manufacturing that are currently overlooked by most government and industrial R&D efforts. Specifically, polymer packaging is an unrealized opportunity that has been underfunded in both the semiconductor and PV industries.

To reach grid parity with traditional energy resources, such as coal and oil, PV modules must be durable and inexpensive. Polymer packaging protects the fragile solar cells from the harsh environmental elements of snow, sleet, and rain. The performance warranty offered by PV manufacturers is based on the anticipated performance integrity of the polymeric packaging. Currently, the highest expense for most PV manufacturers is packaging PV cells into modules. Therefore, lower-cost polymers and efficient manufacturing techniques are required for PV modules to become a competitive energy resource in the United States. The difficulty is maintaining polymer quality and integrity while decreasing costs. Although this is a substantial challenge, there are opportunities for the PV industry to simultaneously decrease cost and improve performance.

The purpose of this book is to familiarize the reader with current and future opportunities in PV polymeric packaging. The first chapter introduces basic polymeric concepts, and Chapters 2 and 3 detail the requirements and specifications for polymers in commercial PV modules. Chapter 4 describes packaging processing techniques and provides a troubleshooting guide to improve process yield. Chapter 5 examines the economics behind PV manufacturing and details the reasons for the current high costs of polymeric packaging. The final chapter investigates new frontiers for polymers, which can both improve PV module performance and decrease costs.

References

1. Zoi, C. August 6, 2010. Assistant Secretary for Energy Efficiency and Renewable Energy Memorandum of Decision, Subject: Determination of inapplicability (nationwide limited waiver in the public interest) of section 1605 of the Recovery Act of 2009 (the Buy American provision) to EERE-funded projects for incidental and/or ancillary solar photovoltaic (PV) equipment, when this equipment is utilized in solar installations containing domestically manufactured PV cells or modules (panels). Department of Energy, Washington, DC.
2. Dalton, D.H.; Serapio, M.G.; Yoshida, P.G. September 1999. Globalizing Industrial Research and Development. U.S. Department of Commerce, Technology Administration, Office of Technology Policy, Washington, DC.

Acknowledgments

I would like to thank Dr. Todd Menna and Dr. Daniel Donahoe for providing technical and literary edits for various sections of this book. Also, special thanks go to Aurore Simonnet and Robert Thomas for illustrative assistance.

Finally, a long-awaited thank you is extended to my parents for sacrificing some of their dreams so that I could pursue mine. Thanks to my brother for constantly challenging me to do the impossible, or so it seemed at the time. And last, thanks to Aunt Nancy for her continued support, even though all my decisions did not always appear completely rational.

1

Introduction to Polymers

1.1 A Brief Historical Perspective

Early polymer scientists studied natural polymers, such as DNA, RNA, polypeptides, and polysaccharides (e.g., cellulose), but they did not immediately understand how the polymer's chemical structure influenced behavior. For instance, in 1855, Alexander Parkes discovered that heated cellulose could be dissolved in a solvent and molded into various shapes. This modified cellulose was commercially used as an ivory substitute for high-value luxury items, such as billiard balls and pianos [1]. At this time, scientists envisioned hardened cellulose as a complex mass of randomly bonded atoms. When Hermann Staudinger published his theory that polymers were composed of atomic chains, scientists started to understand the true causal link between chemistry and macroscopic properties. Theodore Svedberg validated this theory in 1924 by isolating polymer chains using ultracentrifugation [2]. This discovery is credited as the impetus for the modern age of synthetic polymers.

Once scientists understood polymeric structures, they invented synthetic methods for duplicating the molecular architecture. The majority of synthetic polymers of modern significance were patented and commercialized as part of the World War II effort. Polyethylene (circa 1933), polypropylene (circa 1954), polystyrene (circa 1929), and polyethylene terephthalate (circa 1941) constitute the largest global production of polymers [3–5]. After World War II, polymers were commercialized in the public sector, and their global production experienced exponential growth [6]. In 1950, the annual production of polymers was approximately 3.3 billion pounds, and by 2008 it was 540 billion (Figure 1.1). Only during the recent economic downturn has polymer production slowed.

This initial growth was principally due to the commercialization of polymers for consumer packaging, specifically food packaging. Polymers had higher mechanical and environmental durability than paper but were not as expensive as glass and metal. Today, consumer packaging remains the largest use for polymers.

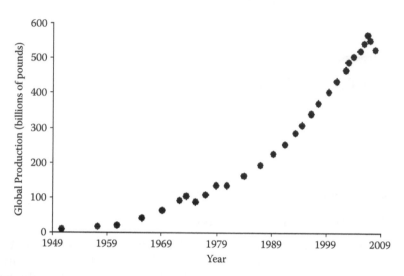

FIGURE 1.1

Global polymer production from 1950 to 2008. (From *Plastics News FYI....*, Global Plastics, Plastics Resin Production over the Years, October 30, 2009, http://plasticsnews.com/fyi-charts/materials.html?id=17004, YGS Group. Used with permission of Plastics News Copyright © 2010. All rights reserved.)

Even though food packaging is the highest-volume application, polymer packaging is used in a number of consumer products. One of the growing consumer applications is photovoltaics (PVs), also generically known as solar, packaging. Since PV's commercialization in the late 1970s, polymers have been proposed as a means of packaging and framing photovoltaic cells. They have received increased interest as the PV market tries to find cheap material choices that will further reduce their manufacturing costs. Chemical manufacturers have responded by marketing polymers for PV applications. However, due to the relative infancy of this application, PV manufacturers have yet to standardize selection criteria and qualification testing.

Most PV manufacturers have a limited polymer science staff; therefore, it is best to review polymer basics before discussing specifics. The following introduction to polymer science is limited to the topics and polymers immediately relevant to the PV packaging requirements covered in later chapters. Here a limited subset of polymeric classifications, behaviors, and processing techniques is included in this discussion with appropriate tabular data.

1.2 Chemical Structure, Nomenclature, and Morphology

A polymer is a large molecular chain with a repeating chemical structure and high molecular weight (Figure 1.2). Polymers are named for the small

FIGURE 1.2
General depiction of a polymerization reaction.

molecules used to synthesize the long chains. Their synthesis is called polymerization. The individual molecules are referred to as monomers before they are polymerized into the polymeric chain, after which they are known as repeat units. The degree of polymerization (P_n) is the number of these repeat units in the chain and is represented in the chemical structure by a subscript; P_n is dimensionless. The degree of polymerization multiplied by the molecular weight of the repeat unit (M_i), in units of grams per mole (g/mol), is the molecular weight of the polymer chain (M), also measured in grams per mole (Equation 1.1):

$$M = P_n M_i \tag{1.1}$$

Each polymer chain is composed of a discrete number of repeat units described by a single molecular weight, but commercial formulations are composed of a number of chains described by a distribution of molecular weights. The weight average molecular weight ($\overline{M}w$) is one method for describing this distribution. The weight average molecular weight is the summation of the product of the number of chains at a specified molecular weight (n_i) and the molecular weight of each chain (M_i) squared divided by the summation of the product of the number of chains at a specified molecular weight and the molecular weight of each chain (Equation 1.2):

$$\overline{M}_w = \frac{\sum_{i=1}^{i=\infty} n_i M_i^2}{\sum_{i=1}^{i=\infty} n_i M_i} \tag{1.2}$$

Although the weight averaged molecular weight is not specified on a product data sheet, the polymer's physical form gives an indication of its size. When a polymer is offered as a solid, the weight averaged molecular weight is high, on the order of a few million. When offered as an oil or grease, the weight averaged molecular weight is typically a few orders of magnitude lower, a few hundreds to thousands of grams per mole. The polymers used for PV applications will be solids with weight averaged molecular weights in the millions.

A generalized chemical structure is used to depict commercial formulation chemistry. Specifically, because each chain has a different degree of

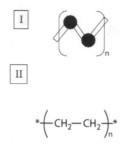

FIGURE 1.3
(I) General depiction of a homopolymer and (II) a specific example of polyethylene.

polymerization, a letter subscript rather than a numerical value is used to denote a distribution of chain lengths in the formulation.

The chain's chemical structure is included in the nomenclature. The polymer is known as a homopolymer when the same repeat unit occurs throughout the length of the chain a number (n) of times (Figure 1.3). For homopolymers, the name of the monomer, or repeating structural unit, makes up the root of the word. The prefix *poly-* indicates that monomer has been synthesized into a polymer. For instance, polyethylene is a polymer composed from ethylene monomers (Table 1.1).

When chemically different repeat units are linked into a polymer chain, it is classified as a copolymer. Generally, copolymers are named after the two monomers constituting the polymer chain with the word *copolymer* at the end of the phrase. Alternatively, the names of the two monomers, or structural units, can be preceded with the prefix *poly-* and connected with the phrase *-co-*. For example, ethylene vinyl acetate copolymer (e.g., polyethylene-*co*-vinyl acetate) is composed of an ethylene monomer polymerized with a vinyl acetate monomer (Figure 1.4). The different subscripts n and m indicate that the number of incidences of these two repeat units are not equivalent. Again, the exact values are dependent on the polymer's molecular weight.

In these previous examples, the specific copolymer type has not been identified in the name. It is common for manufacturers to generically specify the copolymer as a means to conceal their proprietary formulation. However, the reader should be aware that there are multiple classifications of copolymers. Common classifications include statistical, alternating, random, graft, and block. Alternating and block copolymers are specifically relevant for PV applications and will be the focus of the discussion.

Statistical copolymers incorporate repeat units that follow a statistical pattern. Random and alternating copolymers are a subclassification of statistical copolymers. Random copolymers have repeat units scattered along the polymer chain with no specified pattern. The polymer is named with the prefix *poly-* and the two monomer names separated by the phrase *-ran-*. An

TABLE 1.1

Homopolymer Names with Corresponding Monomer and Polymer Structure

Polymer Name	Monomer Structure	Polymer Structure
Polyethylene terephthalate	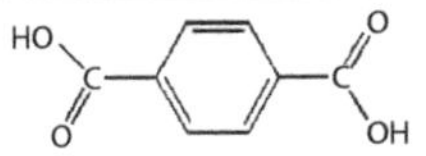 terephthalic acid  ethylene glycol	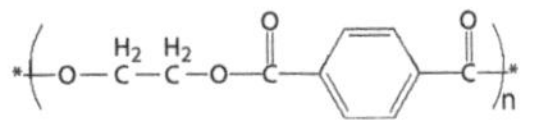
Polyethylene	ethylene	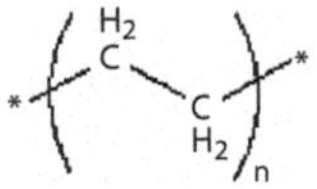
Polypropylene	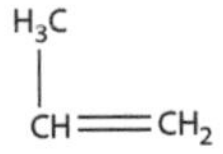 propylene	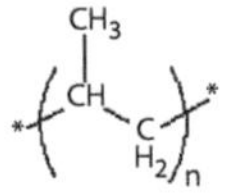
Polystyrene	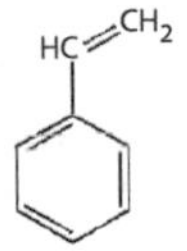 styrene	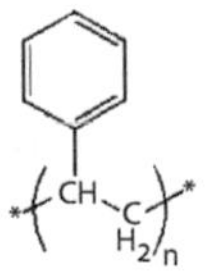

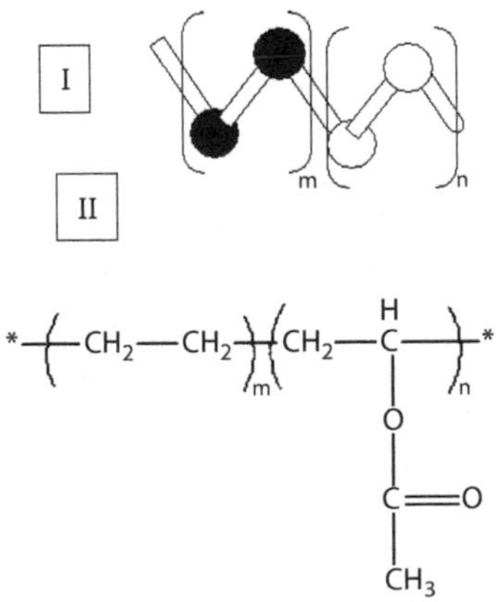

FIGURE 1.4

(I) General depiction of a copolymer and (II) a specific example of ethylene vinyl acetate copolymer.

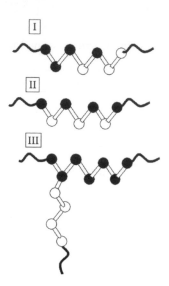

FIGURE 1.5
General depiction of (I) random, (II) alternating, and (III) graft copolymers.

alternating copolymer is composed of two or more repeat units character-ized by an alternating frequency along the chain (Figure 1.5). The polymer nomenclature follows the same pattern as specified above, except the phrase *-alt-* will separate the names of two monomers, or structural units.

A graft copolymer has one homopolymer composing the backbone and another polymer dangling off the side. The nomenclature is to name each of the polymers separately and combine the two names with the phrase *-graft-* or *-g-*.

A block copolymer has two or more segments of the polymer chain with different repeat units composing each segment (Figure 1.6). In this case, each polymer is named separately and linked together with a *-b-* or *-block-* to des-ignate that the two polymers form one chain. For example, polyethylene-*b*-polymethylacrylic acid salt-*b*-polymethylacrylate is a polymer chain of polyethylene linked to a salt of polymethylacrylate linked to chain of polym-ethylacrylate. Again, the number of incidences of each of these repeat units is arbitrarily represented as *m, n,* and *x* to indicate a distribution of chain lengths in the formulation.

There are a number of block copolymer architectures not identified in the nomenclature. For instance, it is also possible to have two polymers linked linearly with one junction point or linked into a circle with two junction points (Figure 1.7). When block copolymers include three or more polymer chains, they can link together to form star, linear, and circular architec-tures. Using a triblock polymer as an example, these architectures result in one, two, and three junction points, respectively. The structure depicted in Figure 1.6 is a linear architecture with two junction points.

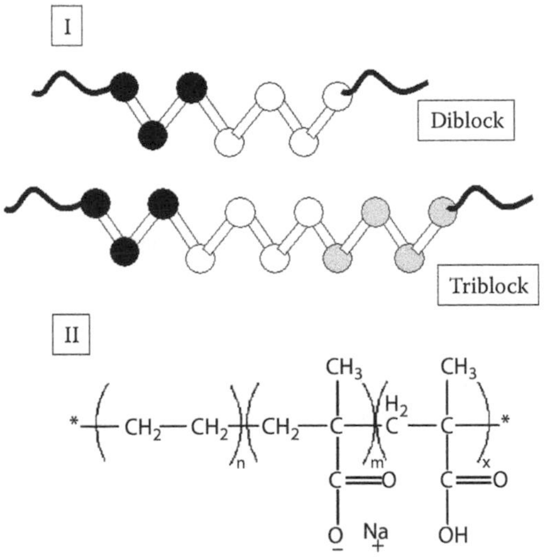

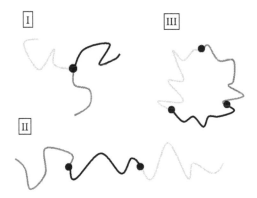

FIGURE 1.6

(I) General depiction of block copolymers and (II) a specific example of polyethylene-*b*-polymethacrylic acid salt-*b*-polymethylacrylate copolymer.

FIGURE 1.7

Block copolymer architectures (I) star, (II) linear, and (III) circular.

For simplicity, polymer scientists will condense polymer nomenclature into a two- to three-letter abbreviation. You may be familiar with these abbreviations if you have recently turned over a plastic bottle. In 1988, due to the escalating use of plastics for disposable consumer packaging, the plastics industry devised recycling logos to insure plastic products could be easily separated after disposal. The abbreviation comes from a combination of the letters used in the polymer name. A list of the most relevant is provided in

TABLE 1.2

Recycling Codes of Polymer Packaging

Identification Code	Polymer Name	Abbreviation
1	Polyethylene terephthalate	PET, PETE
2	High-density polyethylene	HDPE
3	Polyvinyl chloride	PVC
4	Low-density polyethylene	LDPE
5	Polypropylene	PP
6	Polystyrene	PS
7	All other polymers	Other

Source: Data from J. Brandrup, E.H. Immergut, E.A. Grulke, 1999, *Polymer Handbook*, 4th Ed., New York: John Wiley & Sons.

TABLE 1.3

Polymer Name, Abbreviation, and Trade Name for Relevant Photovoltaic (PV) Packaging

Polymer Name	Abbreviation	Trade Name
Polyethylene terephthalate	PET or PETE	Rynite®, Mylar®, Melinex®
Ethylene vinyl acetate copolymer	EVA	Elvax®, Escorene™, Ultrathene®, Encapsolar®
Polyvinyl fluoride	PVF	Tedlar®
Polyethylene-*b*-polymethacrylic acid salt-*b*-polymethylacrylate	None	Suryln®
Polydimethylsiloxane	PDMS	Sylgard®

Table 1.2 [9]. For instance, polyethylene terephthalate is commonly referred to as PET or PETE. A complete list of industrial abbreviations for polymers can be found in international standards published by the International Organization for Standardization (ISO). The two relevant standards are ISO 1043-1:2001, "Plastics—Symbols and Abbreviated Terms—Part 1: Basic Polymers and Their Special Characteristics" [7] and ISO 1629-1995, "Rubbers and Latices—Nomenclature" [8].

Each chemical manufacturer creates a trade name to refer to both the polymer and the proprietary additives. When looking up trade names on manufacturers' Web sites, they will often provide the polymer that is included in their formulation but suppress other formulation ingredients. Polymer names, abbreviations, and trade names commonly used in PV applications are provided in Table 1.3. Based on the overlap between Table 1.2 and Table 1.3, some of the polymers used in PV packaging have been adopted from consumer food packaging. The requirements and specifications that necessitate these selections are discussed in Chapters 2 and 3.

Names are derived from their chemical structure, but polymers are classified based on morphology. Chemical interactions between repeat units on the same chains and between different chains cause them to organize into a three-dimensional morphology. The morphology can be described as semicrystalline, amorphous, ionic clusters, or cross-linked. These different morphologies influence the polymer's functional properties.

The size of polymer chains makes their morphology complicated. Small molecules, like sodium chloride (e.g., table salt), form single crystals. However, polymer chains cannot rearrange and organize into a single crystal due to their large chain structure. Instead, a semicrystalline polymer has regions that are organized into crystalline domains (Figure 1.8). These domains are frozen into place and require heat or mechanical force to separate.

In amorphous polymers, there is no defined organization. The polymer chains organize into their equilibrium, random configuration (Figure 1.9). Above a critical molecular weight, these polymeric chains are entangled. However, given enough time and energy, the chains are free to reptate and slide past one another without breaking chemical bonds.

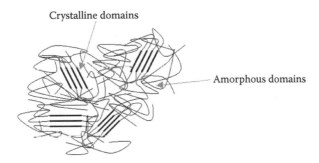

FIGURE 1.8
Depiction of a semicrystalline polymer.

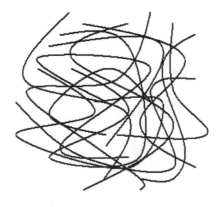

FIGURE 1.9
Depiction of amorphous polymer chains intertwined in a random configuration.

Ionomers contain ionic repeat units in their polymeric structures. The charge on the polymer chain occurs from an ionic group on the repeat unit. These ionic repeat units have a permanent charge that must be neutralized by a counterion, such as a zinc or a sodium cation. The morphology is dictated by the repulsion of adjacent polymer chains, and the requirement that the charges be neutralized when the chains are in their equilibrium configuration.

Various morphologies have been observed based on the counterion and the ionic repeat unit concentration, also termed *percentage of ionization* [10,11]. The intercluster model is specifically relevant for this discussion. The polymers organize into domains with counterions in the center and the ionic repeat units organized on the circumference (Figure 1.10). The ionic agglomerations act as pseudo-cross-links that limit chain movement, because the overall polymer charge must remain neutral.

A cross-linked morphology describes polymer chains chemically bonded together (Figure 1.11). Although many forms exist, the morphology can be

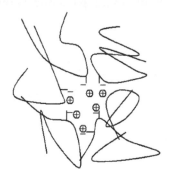

FIGURE 1.10
Depiction of ionic domain in an ionomer morphology.

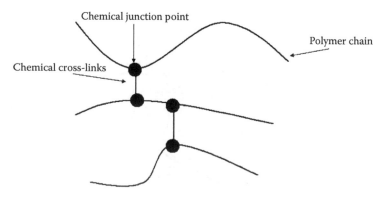

FIGURE 1.11
Depiction of a segment of a cross-linked polymer.

commonly visualized as a brick pattern with two chains chemically bonded to one another with a tie molecule, also called a cross-link. The polymer chains are chemically constrained from sliding past one another by these chemical cross-links.

1.3 Polymeric Classification Based on Thermal and Mechanical Properties

Like most materials, polymers undergo thermal transitions. Almost all polymers exhibit a glass transition temperature, but only ionomers exhibit an order–disorder transition. Below the glass transition, the chains are effectively frozen over short time scales, and macroscopic behavior is described as rigid. Above the glass transition, kinetic energy increases, and the polymers reptate away from each other, increasing the polymers' internal volume. The macroscopic behavior is described as rubbery. Conceptually, the same is true for ionomers, except below, the order–disorder transition chains are ordered into clusters; and above the transition, the clusters dissolve.

Both the glass (T_g) and order–disorder ($T_{order-disorder}$) temperatures are second-order transitions. They are observed as a change in the slope of heat (q) added to a system graphed as a function of system temperature (T) (Figure 1.12). These transitions are defined as the temperature where the two slopes intersect.

Only a subset of polymers exhibits a melt temperature (T_m). A polymer must have crystalline domains to exhibit a melt temperature. It is a first-order transition denoted by a discontinuity in the heating curve. As the polymer absorbs heat, the crystalline domains melt. Variations in the polymer chain length create crystalline imperfections. This causes the crystalline domains to melt over a temperature range, rather than at a discrete temperature. As the crystal domains melt, all the heat is consumed in the phase transition. Therefore, there is no change in polymer temperature during the transition (Figure 1.13).

Each polymer classification has a characteristic thermal transition (Table 1.4) [12,13]. Amorphous polymers flow into a new shape at temperatures above the glass transition (T_g). A semicrystalline polymer will exhibit a glass transition temperature but will not flow until the applied temperature exceeds the melting temperature (T_m). Ionomers have an order–disorder transition temperature that must be exceeded to achieve homogenous flow of the polymeric chains. A cross-linked polymer will never irreversibly flow, because the chemical cross-links prevent chain movement.

These transition temperatures are typically not included in the product data sheets of commercial polymers. It is expected the consumer is aware

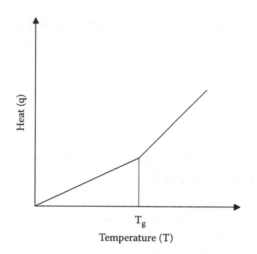

FIGURE 1.12
Glass transition temperature depicted on a heat versus temperature curve.

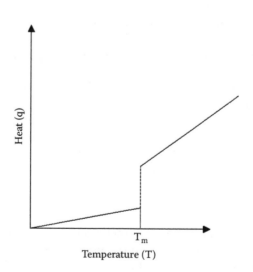

FIGURE 1.13
Melt temperature depicted on a heat versus temperature curve.

TABLE 1.4

Polymer Classification, Polymer Name, and the Corresponding Thermal Transitions

Classification	Polymer Name	$T_{order-disorder}$ (K)	T_g (K)	T_m (K)
Thermoplastic	Polyethylene terephthalate	N/A	388–342	538
	Ethylene vinyl acetate copolymer	N/A	235–231	379–318
	Polyvinyl fluoride	N/A	337	466–463
Thermoset	Epoxy	N/A	N/A	N/A
Ionomer	Polyethylene-*b*-polymethacrylic acid salt-*b*-polymethylacrylate	331	148	373
Elastomer	Polydimethylsiloxane	N/A	150	N/A

Source: Data from J.E. Mark, 1999, *Polymer Data Handbook*, Oxford: Oxford University Press; H.F. Mark, 1985, *Encyclopedia of Polymer Science and Engineering*, 2nd Ed., Vol. 15, New York: Wiley.

of these inherent polymer properties. However, polymer manufacturers will typically specify a recommended material processing condition within the data sheet. Although specific temperatures and corresponding polymer properties are not itemized, these recommendations are based on the aforementioned polymeric transitions.

Mechanical behavior is typically assessed in either tensile or compression mode. The stress is graphed as a function of strain for both analyses.

Stress is a tensor quantity that describes the vector force exerted on various spatial points in the sample. For simplicity, in this text, uniaxial deformation will be assumed. Under these conditions, the tensor is equivalent to a scalar quantity.

The engineering stress (σ_e) can be approximated as the normal force (F_n) divided by the initial area acted upon (A_o). Stress is measured in units of newtons per square meter (N/m^2), also known as a pascal (Pa) (Equation 1.3). For polymers, it is common to find stress units reported as Megapascal (1 MPa = 10^6 Pa).

$$\sigma_e = \frac{F_n}{A_o} \tag{1.3}$$

Strain is defined as a change in shape due to the presence of stress. The strain, under uniaxial stress, can be defined as engineering strain when a test specimen has a larger length than the cross-sectional area (Equation 1.4). Engineering strain (ε_e) is the change in specimen length ($\delta = L_f - L_o$) divided by initial length (L_o). Strain is a dimensionless quantity defined as a ratio or percentage.

$$\varepsilon_e = \frac{\delta}{L_o} \tag{1.4}$$

If the polymer breaks under an applied stress, the strain at break (ε_b) is the total percent change in length.

Hooke's law is a common mathematical expression used to describe the relationship between stress and strain. Hooke's law states that stress is linearly proportional to strain. It is an empirical observation that only applies at low strain values. When Hooke's law is observed, Young's modulus (E) is defined as stress divided by strain (Equation 1.5):

$$E = \frac{stress}{strain} \tag{1.5}$$

Some of these mechanical properties can be found on product data sheets provided by polymer manufacturers. The mechanical behavior listed will depend on the supplier's anticipated application for the polymer. In most cases, an exhaustive list of mechanical behavior has been made to characterize new commercial formulations. Therefore, if the desired properties are not on the data sheet, they often can be provided upon request. Alternatively, there are a number of polymeric books and resources cited in this chapter's reference section that can be consulted for these values [13–16].

Each polymeric classification has a different expected mechanical behavior (Table 1.5, Figure 1.14). Mechanical behavior is dependent on the chemical structure and the measurement temperature relative to the inherent polymeric transitions just described. Because some manufacturers specialize

TABLE 1.5

Polymer Classification, Polymer Name, Corresponding Tensile Modulus, and the Elongation to Break

Classification	Polymer Name	Tensile Modulus (MPa)	Elongation to Break (%)
Thermoplastic	Polyethylene terephthalate	1700	180
	Ethylene vinyl acetate copolymer	3138–2062	850–675
	Polyvinyl fluoride	2300–1170	175–100
Thermoset	Epoxy	28,000–3000	85–1
Ionomer	Polyethylene-b-polymethacrylic acid salt-b-polymethylacrylate	290–280	500–470
Elastomer	Polydimethylsiloxane	200	725–430

Sources: Data from J.E. Mark, 1999, *Polymer Data Handbook,* Oxford: Oxford University Press; H.F. Mark, 1985, *Encyclopedia of Polymer Science and Engineering,* 2nd Ed., Vol. 15, New York: Wiley.

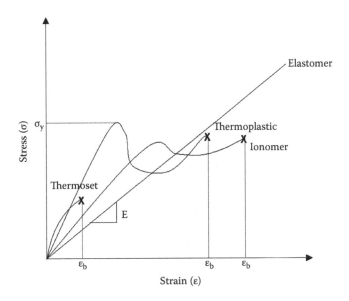

FIGURE 1.14
The typical mechanical behavior of the major classifications of polymers.

TABLE 1.6

Polymer Classification and Some Manufacturers

Classification	Manufacturers
Thermoplastics	Exxon, Total, DuPont
Thermosets	Henkel, Epoxy Technology
Ionomers	DuPont
Elastomers	Dow Corning, Momentive, NuSil Technology

in a specific polymeric class, it is assumed the design engineer recognizes these differences in mechanical response prior to approaching the manufacturer for pricing and samples (Table 1.6). However, to provide a baseline knowledge for readers, the mechanical response for each of these classifications is explained below.

1.3.1 Thermoplastics

Thermoplastics are typically sold as one solid component that flows only when heated. Thermoplastics generally have a glass transition (T_g) and may or may not have a melt temperature (T_m). Commonly, at least one, if not both, of the transitions is above ambient, and the polymer will be rigid at room temperature.

To counter this inherent thermal property, manufacturers add small organic molecules, called plasticizers, to some formulations. Phthalates are the most common classification of plasticizers. They are often used in polyvinyl chloride (e.g., PVC) polymers, and they lower the glass transition (T_g) of the polymer, making it pliable at room temperature. This is an important distinction in the chemistry of thermoplastics.

Phthalates have been around for decades, and they are the cause of a "new car smell." Historically, in this context, the presence of phthalates and their diffusion into the air creates consumer satisfaction. In fact, the smell of phthalates was once so popular that it was sold at car washes as air freshener. However, in recent years, these same chemicals have come under increased environmental and health scrutiny. For instance, recent studies by the National Institutes of Health (NIH) have shown that small amounts of phthalates are released from plastic bottles used to packaged soda and water [17]. High quantities of phthalates have been associated with health risks such as liver disease and certain types of cancer. Research is continuing to assess these risks of thermoplastic packaging [18,19]. Small molecule migration from thermoplastics used in PV packaging will be a theme revisited in various chapters. In this case, the emphasis is not on health risks to PV consumers but on the failure mechanisms caused by small molecule migration.

Despite the ability to manipulate the glass transition of thermoplastics, most have an above ambient melt temperature. The crystalline domains at room temperature create mechanical strength. For this reason, a large tensile force must be applied to the polymer in order to induce deformation. Above the yield stress (σ_y), the polymer will undergo an irreversible strain called plastic deformation (Figure 1.15). The region before the yield is defined as the elastic region, and the region after deformation is called strain softening. When stressed beyond the yield point, the macroscopic sample will not return to its original shape. During cold drawing, there is no change in stress with further polymer elongation. Finally, the polymer breaks after strain hardening.

Ethylene vinyl acetate copolymer, polyvinyl fluoride, and polyethylene terephthalate have above ambient melt transitions. During processing, the polymers are heated above their melt temperature so they can flow into their final shape. If the polymer temperature is quenched back to room temperature after processing, the polymer does not have time to reorganize into its equilibrium configuration, and the crystalline domains will not reform. The absence of crystalline domains significantly alters the expected material properties. For this reason, it is important to test the critical mechanical properties after processing.

1.3.2 Thermosets

Thermosets are generally sold as two components, a base and a catalyst, with multiple processing options. Dispensed molding is the most relevant

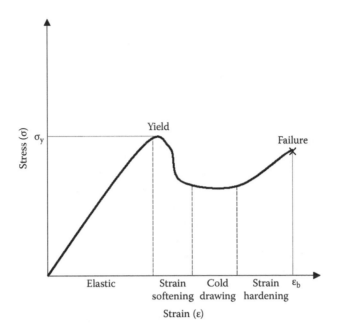

FIGURE 1.15
Behavioral regions on the stress–strain curve of a thermoplastic under tension.

processing technique for this discussion. Dispensed thermosets are separately packaged and dispensed directly into the product. Once mixed, the base reacts with the catalyst to form a three-dimensional cross-linked network. Alternatively, some thermosets can be compression molded into the desired part. In this case, heat acts to initiate the reaction to cross-link the polymer chains.

A high cross-link density, the number of cross-links per unit area, defines thermoset morphology. It is this chemical structure that leads to their characteristic mechanical behavior. Thermosets have the highest modulus and lowest elongation to break of all the cited classifications (Figure 1.14, Table 1.4). A large modulus means when undergoing deformation an excessive amount of force must be applied before there is a change in strain. Excessive force is required to break the chemical cross-links and allow the polymeric chains to elongate. After enough bonds break, there will be permanent macroscopic deformation. Due to the high cross-link density, the macroscopic sample will not elongate very far before it breaks.

The cross-linked network has no melt temperature and will not flow into a different shape when heated. For instance, the epoxy, cited in Table 1.4, has no defined thermal transition (T_g or T_m). Instead, upon excessive heating, the epoxy will chemically decompose before flowing. Once decomposition has occurred, the thermoset will not reform its original shape.

1.3.3 Ionomers

Ionomers are typically block or graft copolymers. Each segment in the chain has specific attributes that contribute to the copolymer's properties. One segment is typically semicrystalline, and one is amorphous. The ionomer will have both semicrystalline and amorphous regions, and it will be characterized by both melt and glass transition temperatures.

The amorphous segments typically contain acidic groups susceptible to ionization, a chemical process leading to the formation of ionic charges along the backbone of the polymer chain. When these segments are ionized, the thermal signature becomes more complicated. The ionomer now has amorphous, crystalline, and ionic clusters in its morphology. Therefore, the polymer can exhibit three different types of transitions (T_g, T_m, and $T_{disorder-order}$). The highest temperature transition dictates the maximum temperature required for the polymer to flow. As an example in Table 1.4, polyethylene-*b*-polymethacrylic acid salt-*b*-polymethylacrylate has all three transitions, and the melt temperature is the highest. Above 373 K, polyethylene-*b*-polymethacrylic acid salt-*b*-polymethylacrylate can be molded into a new shape.

Ionomers are sold under various chemical grades in sheet and pellet form. Each grade is characterized by the percentage of ionization and the counterion used to balance the charge. These differences in chemical structure create slightly different mechanical, thermal, and weathering properties among the various grades. Therefore, the relative placement of the ionomer to thermoplastics depicted in Figure 1.14 is highly dependent on the two polymers compared. The effect of chemical structure on macroscopic properties will be discussed in detail for polyethylene-*b*-polyacrylic acid salt-*b*-polyacrylate in Chapter 3.

1.3.4 Elastomers

Elastomers are sold as either one- or two-part chemistries. A one-part chemistry is purchased as a single package or canister from the manufacturer. The most common is called condensation chemistry. The manufacturer provides the base and catalyst in one canister, but the reaction does not occur until a coreactant, typically water, is absorbed from the surrounding environment. The three components react to form a cross-linked structure. In contrast, a two-part chemistry is purchased as two separate packages from the manufacturer. One package contains the base and the second the catalyst. The two components must be mixed in a defined ratio for the elastomer to properly cross-link.

The subambient glass transition and low cross-link density of elastomers impart the rubbery mechanical properties associated with this class. A subambient glass transition means the polymer chains have significant motion inhibited only by the chemical cross-links. The low concentration of chemical cross-links per unit area allows elastomers to elongate under minimal stress without chemical bond rupture. This gives elastomers a low modulus and high elongation to break. Polydimethylsiloxane has the lowest reported modulus of the polymers referenced in Table 1.5. In addition, elastomers can be stretched to over 100% of their initial length and return to their original shape without permanent deformation. For example, polydimethylsiloxanes have the largest elongation to break, 430% to 725%, of the various polymers discussed. Due to their cross-linked chains, elastomers cannot flow into a new shape when heated. Like thermosets, extreme heat will cause decomposition prior to flow.

References

1. Worden, E.C. 1911. *Nitrocellulose Industry*. New York: Van Nostrand.
2. Holde, K.E.; Hansen, J.C. 1998. Analytical Ultracentrifugation from 1924 to the Present: A Remarkable History. *Biochemistry and Molecular Biology* 11: 933–943.
3. Scheirs, J.; Long, T.E. 2003. *Modern Polyester: Chemistry and Technology of Polyesters and Copolyesters*. New York: John Wiley & Sons.
4. Martin, H. 2007. *Polymers, Patents, Profits: A Classic Case Study for Patent Infighting*. Weinheim: Wiley-VCH.
5. Pohlemann, H.G.; Echte, Q. 1981. *Polymer Science Overview*. Washington, DC: American Chemical Society.
6. Sailors, H.R.; Hogan, J.P. 1981. History of Polyolefins. *Journal of Macromolecular Science, Part A* 15:1377–1402.
7. ISO 1043-1:2001, "Plastics—Symbols and Abbreviated Terms—Part 1: Basic Polymers and Their Special Characteristics" International Organization for Standardization. Geneva, Switzerland. http://www.iso.org
8. ISO 1629-1995, "Rubbers and Latices—Nomenclature" International Organization for Standardization. Geneva, Switzerland. http://www.iso.org
9. Brandrup, J.; Immergut, E.H.; Grulke, E.A. 1999. *Polymer Handbook*, 4th Ed. New York: John Wiley & Sons.
10. Holden, G.; Kricheldorf, H.R.; Quirk, R.P. 2004. *Thermoplastic Elastomers*, 3rd Ed. Munich: Hanser.
11. Williams, C.E.; Russell, T.P.; Jerome, R.; Horrion, J. 1986. Ionic aggregation in model ionomers. *Macromolecules* 19: 2877–2884.
12. Mark, J.E. 1999. *Polymer Data Handbook*. Oxford: Oxford University Press.
13. Mark, H.F. 1985. *Encyclopedia of Polymer Science and Engineering*, 2nd Ed., Vol. 15. New York: Wiley.
14. Harper, C.A. 1975. *Handbook of Plastics and Elastomers*. New York: McGraw-Hill.
15. National Institutes of Health. 2000. NTP-CERHE Expert Panel Report on Di N octyl phthalate. http://cerhr.niehs.nih.gov/chemicals/phthalates/dnop/DnOP-final-inprog.PDF

16. Mark, J.E. 2007. *Physical Properties of Polymer Handbook*, 2nd Ed. New York: Springer.
17. National Institutes of Health. 2000. NTP-CERHE Expert Panel Report on Di N octyl phthalate. http://cerhr.neihs.nih.gov/chemicals/phthalates/dnop/DnOP-final-inprog.PDF
18. Halden, R.U. 2010. Plastics and Health Risk. *Annual Review of Public Health* 31:179.
19. Autian, J. 1973. Toxicity and health threats of phthalate esters: Review of the literature. *Environ Health Perspect.* 4:3–26.

2

Certification and Characterization
of Photovoltaic Packaging

2.1 Overview of Photovoltaic Installations

Although there are other forms of solar energy (e.g., solar thermal), the topics covered in this book will be limited to those relevant to photovoltaic energy. Photovoltaic (PV) installations are divided into three subassemblies capable of energy generation: the cell, the module, and the array.

A PV cell is the smallest subassembly capable of producing power. The active layer is composed of a semiconductor film responsible for converting light into electricity. This phenomenon is called the photoelectric effect, and it gives this branch of solar energy its name. For simplicity, p-n–type semiconductors, such as single crystalline silicon, will be used as an example. Incident light with energy equal to or greater than that of the semiconductor's band gap (E_g) will knock electrons out of their atomic orbitals and into the n-type semiconductor layer, also known as the electron donor. This leaves behind a hole in the p-type semiconductor layer, also termed the electron acceptor (Figure 2.1). The band gap is an inherent characteristic of the semiconductor material, and it makes each chemistry sensitive to specific wavelengths of light. The intermediate zone where these two layers meet is called the p-n junction layer. The separation of charges creates a measurable voltage. Connecting the grid metallization and metallic backsheet through the interconnects allows for the flow of electrons out of the n-type layer and to the p-type layer. The flow of electrons through the circuit results in a measurable current. The voltage multiplied by the current will define the solar cell's power, and a few milliwatts (mW) is typical.

A collection of fragile PV cells is electrically connected into a string, and a series of strings are packaged into a module. The cells are composed of a delicate chemistry that cannot be directly exposed to environmental elements. Many of the semiconductors degrade in the presence of moisture and easily fracture when exposed to sleet or hail. Most importantly, the bare cells are an electrical hazard if touched while irradiated with sunlight. For these reasons, PV cells are packaged with multiple layers of glass and polymers to form the

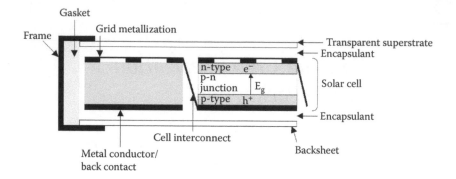

FIGURE 2.1
Cross section of a typical single crystalline silicon photovoltaic (PV) module.

finished product, called a module. The primary barrier layers are a glass superstrate and polymeric backsheet. The solar cell is embedded in an intermediate layer, called the polymer encapsulant. All layers are encompassed in a frame and held into position with an elastomeric gasket or adhesive. The module is commonly the smallest consumer unit a PV manufacturer sells. Some manufacturers use the terms *module* and *panel* interchangeably, but the term *module* is slightly more common. The power produced by a module can vary between 70 and 250 peak watts (W_p), depending on module size and cell chemistry.

A series of electrically connected models create a PV array. The consumer's energy requirements define the array footprint; these calculations are presented in Chapter 5. A typical household would purchase a few kilowatts of power. This would require 10 to 20 modules. An industrial installation is hundreds of thousands of kilowatts to a few Megawatts (MW). This power requirement necessitates hundreds to thousands of modules.

2.2 Selection Requirements for Photovoltaic Packaging

PV companies are diversified based on the type of cell they manufacture, and each manufacturer usually specializes in one cell chemistry. A PV cell is typically formed from one of the following chemistries: amorphous silicon (a-Si), polycrystalline silicon (p-Si), crystalline silicon (c-Si), cadmium telluride (CdTe), copper indium diselenide (CIS), or copper indium gallium diselenide (CIGS). Despite this specialization, most PV manufacturing is divided into two distinct units: the front end produces the cell, and the back end is responsible for packaging the cells into modules. This book focuses on polymer applications in the back end of PV manufacturing.

TABLE 2.1

Photovoltaic Applications, Commonly Used Polymers, and Selected Trade Names

Polymer Name	Trade Name	PV Packaging	Balance of System (BOS) Application
Polyethylene terephthalate	Rynite®, Mylar®, Melinex®	Backsheets	Frames, junction box enclosure
Ethylene vinyl acetate copolymer	Elvax®, Encapsolar®, Escorene™, Ultrathene®	Encapsulants	None
Polyvinyl fluoride	Tedlar®	Backsheets	None
Polyethylene-*b*-polymethacrylic acid salt-*b*-polymethylacrylate	Suryln®	Encapsulants	None
Polydimethylsiloxane	Sylgard®	Encapsulants	Frame adhesives, junction box pottants

Polymers are heavily used in both module packaging and the balance of system (BOS) (Table 2.1). Applications in packaging include encapsulants, backsheets, and adhesives (Figure 2.2). The superstrate is commonly composed of float glass, but some researchers have attempted to use polymers for this application. All of these components are used to seal the PV cells into modules. Module configuration is differentiated based on the relative placement between the PV cells and these components. Various manufacturers have commercialized superstrate-bonded, substrate-bonded, and laminated modules [1,2]. Today, laminated modules are the most common configuration. The balance of systems includes all the components required for fixturing and electrically connecting the modules to a home or business. It includes but is not limited to the frames, junction boxes, and batteries. There is some overlap between the type of polymers used in packaging and BOS applications. Despite this overlap, each application has different material requirements (Table 2.2).

Because the encapsulants and backsheets encase the PV cells, they directly influence the product's performance and have the highest material requirements. Encapsulants and backsheets require simultaneous consideration of optical, weathering, flammability, thermal, electrical, and mechanical properties during material selection. They act as an optical coupling agent, environmental barrier, heat sink, electrical insulator, and structural component. Each of these functions has specialized material requirements. As an optical coupling agent, the polymers must efficiently guide light into the PV cell. As an environmental barrier, both the encapsulant and backsheet must protect the PV cell from snow, ice, rain, and soil. As a heat sink and electrical insulator, they must simultaneously draw heat from the PV cell and provide electrical insulation from the consumer. As a structural member, they must absorb shock to prevent mechanical stress from harming the embedded PV cells.

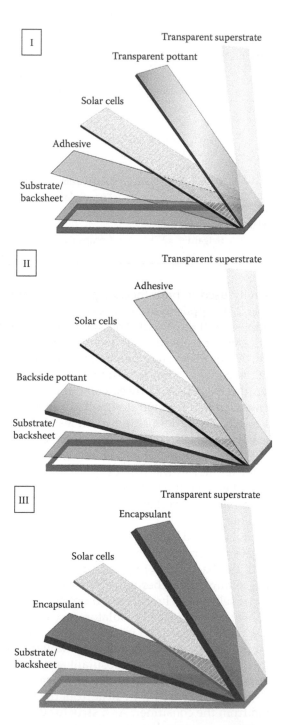

FIGURE 2.2
(I) Substrate bonded, (II) superstrate bonded, and (III) laminated photovoltaic (PV) modules.

TABLE 2.2

Material Properties of Concern to the Design Engineer for Each Photovoltaic Packaging Application

Material Property	Encapsulants	Adhesives	Frames	Backsheets	Junction Box— Enclosure	Junction Box— Pottant
Optical	•			•		
Thermal	•	•		•	•	•
Mechanical	•	•	•	•	•	
Electrical	•			•	•	•
Flammability	•	•		•	•	•
Weathering	•	•	•	•	•	

In contrast, BOS components have fewer material properties that must be evaluated. Individual polymeric BOS components are typically assembled together to perform all the functions required from a single packaging material. For instance, the junction box pottant simultaneously removes heat from the embedded electronics and provides electrical insulation. The junction box enclosure protects the encased electronics from mechanical and environmental forces. The evaluation procedure of these two components combined is similar to that for a single PV packaging component.

The topics outlined in this chapter are meant to be accessible to the design engineers responsible for selecting and designing polymeric components for PV manufacturing. The tests performed to evaluate a polymer's properties are described in the following sections. The material specifications for each property requirement are reviewed in Chapter 3.

These engineering requirements are only a third of the necessary qualification process. These technical concerns need to be balanced with sales and manufacturing requirements. Specifically, customers expect the modules to pass various certifications and compliance testing. These certifications often require a demonstration of product uniformity during manufacturing.

2.2.1 Certification and Compliance Criteria

Polymers have been used for packaging consumer products for decades. The electronics, pharmaceuticals, and food industries all use polymeric packaging. Polymers have been embraced by these industries because they are inexpensive and durable. For these same reasons, polymers have been a favorite for PV cell packaging. Polymeric packaging allows PV manufacturers to lower their offering price, making them more competitive with other energy sources.

In addition to cost, customers are sensitive to product certifications. Manufacturers interested in selling on the global market must provide both environmental and safety certifications. The PV module must be composed of environmentally friendly and recyclable materials as defined by various European directives (i.e., Restriction of Hazardous Substances [RoHS] and the Waste Electrical and Electronic Equipment [WEEE]). In addition, certifications from the Underwriters Laboratories (UL) and the International Electrotechnical Commission (IEC) are a requirement for distribution.

2.2.1.1 Restriction of Hazardous Substances (RoHS) and Waste Electrical and Electronic Equipment (WEEE) Directives

On July 1, 2006, the RoHS directive, also known as Directive 2002/95/EC, came into effect and was an enforceable law in each state of the European Union (EU). Currently, RoHS has only limited the material selection process for electronics. Electronics manufacturers interested in selling in the EU have to insure the materials included in the product are free from the chemical elements of lead (Pb), mercury (Hg), cadmium (Cd), and hexavalent chromium (Cr^{6+}). The organic flame retardants, polybrominated biphenyls (PBBs) and polybrominated diphenyl ether (PBDE), are also excluded by the directive.

All these materials must not be detected over a specified threshold. In this case, the threshold is a weight percent calculation represented as the amount of restricted compound divided by the weight of the homogeneous component. A homogeneous component is a part of the assembled product that can be mechanically separated in a single piece. If any single homogeneous component is tested and found to be above the specified limits, then the entire product is deemed noncompliant. The limit for all elements and flame retardants is currently 0.1 wt%, or 1000 parts per million (ppm). The exception is hexavalent chromium only allowable at the lower concentration of 0.01 wt%, or 100 ppm [3].

Currently, PV modules are exempt from RoHS compliance because they are viewed as a green product that offsets the toxic emissions of other carbon-based energy. In late 2009, the EU opened a debate to increase restrictions on hazardous chemicals by eliminating this exemption [4]. Although the exemption currently remains in effect, alternatives to gain compliance are becoming a larger corporate initiative for PV companies. Thin-film semiconductor materials, specifically cadmium telluride (CdTe), copper indium diselenide (CIS), and copper indium gallium diselenide (CIGS), have restricted elements used in their cell's construction. The PV cell is the principal component in the module responsible for electricity generation; as a result, it is not easily modified to gain compliance. Therefore, removal of these exemptions threatens the survival of these companies by removing the EU from their customer base.

The polymeric components used in the packaging must not contain restricted flame retardants. Different flame retardants can alter a polymer's physical properties. To avoid lengthy requalification for RoHS-compliant polymers, it is easier to insure compliance now in preparation for the inevitable inclusion of PV modules in RoHS. Most polymer manufacturers have started to remove these restricted compounds from their formulation for their electronics customers. During the selection process, the design engineer simply needs to obtain the polymer's certificate of RoHS compliance from the vendor as a first step toward verifying compliance. An independent compliance test is also encouraged.

The WEEE Directive (2002/96/EC) became European law on August 13, 2005. It requires the company to provide its customers with a cradle-to-grave understanding of the manufacturing processes. Part of that understanding includes a take-back program to recycle and properly dispose of hazardous materials in the product [5]. WEEE-compliant products will possess a recycling sticker depicting a wheelie bin. Even though PV is currently exempt from WEEE, most PV manufacturers assume environmental responsibility by providing an annuity for the return of modules at the end of life. In addition, PV Cycle is an industry-wide voluntary take-back program scheduled for complete implementation in the EU by 2015. Ideally, all materials will be recyclable after they are reclaimed. Important to the discussion of this book, those polymers used in the PV packaging should be recyclable.

2.2.1.2 Underwriters Laboratories

The Underwriters Laboratories (UL) is an independent third-party safety certification board for consumer products and materials. Founded in 1894 by William Merrill, an electrical engineer, its history and focus are in the disciplines of thermal and safety sciences. Within the past decade, the company has expanded. Currently, it has customers in 102 countries and world recognition for its certification processes and logo. UL is currently composed of two business units, a nonprofit business chartered to perform fire and safety research to aid in the development of standards, and a for-profit certification board [6,7]. U.S. consumers are familiar with both the U.S. and Canadian logos found on most home electronics. Although this has become a symbol of safety compliance, there is little consumer understanding for the manufacturing requirements for the certification process. The product or material must meet the requirements specified in UL standards to receive a compliance rating. In addition, the company must demonstrate good manufacturing practices during unannounced inspections to confirm a consistent ability to manufacture a compliant product.

In 1941, UL began publishing tests to monitor the fire safety of synthetic polymers. In the following decades as these materials were integrated into consumer products, the number of UL standards designed for polymeric materials increased. UL has developed plastic and polymer test standards for nearly seven decades [6]. Choosing materials that meet UL requirements and have the necessary certifications can decrease the time to market.

A design engineer can narrow a material selection by verifying the material is UL compliant before building a prototype. There are two databases PV manufacturers can use to verify compliance: the UL iQ™ and the UL Recognized Components database. Membership to UL grants access to both databases. The UL iQ database contains 60,000 grades of UL-certified plastics used in a variety of industries. The UL Recognized Components database contains certified polymers specifically for PV packaging. The flammability, weathering, water diffusivity, and thermal properties of commercial resins are maintained within the database.

2.2.1.3 International Electrotechnical Commission

The International Electrotechnical Commission (IEC) was established in 1906. IEC is an international organization composed of technical committees responsible for developing homogeneity in the electrical industry through the development of nomenclature, symbols, ratings, standards, and certifications [8].

Both IEC and UL standards are used in the PV industry. The IEC standard 61730, "Photovoltaic (PV) module safety qualification" [9,10] is of particular importance because it provides definitions for polymeric components based on their operation either as an enclosure or as a support for live electrical parts, an outer surface, or a barrier. The thermal and flammability ratings enforced by UL are dependent on the polymer's operational category. The specifications outlined in Chapter 3 are based on the polymer's classification using these definitions.

2.2.1.4 American Standard for Testing Materials

In 1898, the American Standard for Testing Materials (ASTM) was formed to address a string of railroad failures linked to quality issues with the steel used in railroad ties. A group of scientists gathered to develop a test method to identify the material property causing the failure and to standardize the compositional analysis to prevent future failures [11].

This general mission statement and methodology has endured throughout the decades, although the vision has broadened to include additional disciplines. Today, ASTM continues its work as a nonprofit organization composed of volunteer committee members. ASTM develops Standard Specifications, Test Methods, Practices, Guides, Classifications, and Terminology, as a means to unify various industries. A certified ASTM document can be identified by the organization's logo.

ASTM Standard Test Methods have been adopted by the polymer industry to characterize material properties. Technical data sheets from suppliers will report polymer properties and the ASTM standard used to derive the reported value. These standards are an excellent method for design engineers to rank performance between various polymers during the selection process.

2.3 Optical Properties

Scientists sometimes describe light as propagating waves. Light waves are conceptually similar to ocean waves (Figure 2.3). Both contain apexes and troughs characterized by a wavelength and frequency. A wavelength (λ) is the distance between repeating points in the wave's shape, such as adjacent apexes or adjacent troughs. It is measured in units of length, typically nanometers (nm). One apex and one trough define a cycle. Frequency (ν) is the number of cycles that pass a point in space, typically measured in cycles per second or hertz (1 Hz = $1s^{-1}$).

When light is described as quantized particles, termed *photons*, their energy (E), measured in joules (J), is equivalent to Planck's constant ($h = 6.626 \times 10^{-34}$ J • s) multiplied by their frequency (Equation 2.1). Frequency is equivalent to the speed of light ($c = 299,792,458$ m/s) divided by the wavelength, measured in meters (m). Therefore, an inverse relationship exists between energy and wavelength. For instance, lower wavelengths have higher energy.

$$E = \nu h = h\frac{c}{\lambda} \tag{2.1}$$

The intensity of the light is measured by spectral irradiance, the energy luminating an area, measured in watts per square meter–nanometer (W/(m² • nm)). The light generated by our Sun is referred to as the extraterrestrial spectrum, and the wavelengths of highest intensity are from 250 to 2750 nanometers (nm) (Figure 2.4). The terrestrial solar spectrum describes the light penetrating our atmosphere. There are a number of terrestrial spectra based on the orientation of the Earth to the Sun, and each orientation is described by a solar zenith angle. The spectrum depicted in Figure 2.4 is the light incident on Earth's crust when the solar zenith angle is 48.19°s, also called the absolute air mass of 1.5 (AM 1.5).

Light is categorized in the electromagnetic spectrum. The spectrum has been divided into sections based on wavelength. This discussion will be

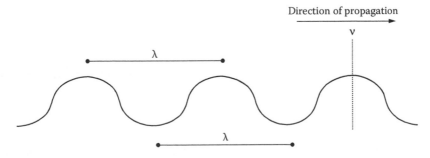

FIGURE 2.3
Depiction of a light wave.

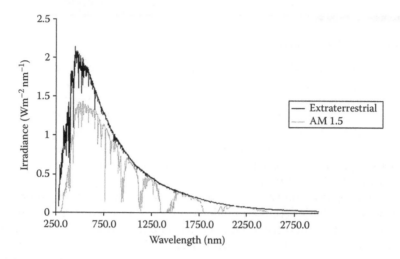

FIGURE 2.4
Extraterrestrial and air mass (AM) 1.5 terrestrial light spectra.

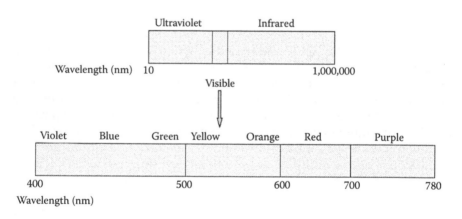

FIGURE 2.5
Selected portion of the electromagnetic spectrum.

confined to the ultraviolet, visible, and infrared portions that compose 9%, 41%, and 50% of the terrestrial spectrum, respectively (Figure 2.5). Most PV cell technologies are highly efficient at harvesting visible light (400 to 780 nm) into electricity.

2.3.1 Material Properties

Encapsulants are the primary elements of PV packaging requiring optical transparency. PV manufacturers set the material selection criteria for encapsulants at 94% to 98% transmission and a degradation of no more than 2%

over 25 to 30 years outdoors. The goal of 98% transmission is based on a 2% maximum loss as a result of a mismatch between the refractive indices of the layered materials.

These goals require the design engineer to consider initial optical transparency and optical durability during service. When performing material selection, design engineers need to obtain the polymer's as received and aged refractive indices. These properties are available from the manufacturer and various polymeric databases [12,13]. Despite these resources, PV engineers often perform their own experiments to verify performance. It is important to note the contrast in how these measurements are performed by polymer and PV manufacturers in order to avoid misuse of information and inaccurate performance predictions.

2.3.1.1 Refractive Index Measurements

Technical data sheets from a polymer manufacturer will typically have a refractive index (n) listed in the property section. The refractive index is the dimensionless ratio of the velocity of light in a vacuum to the velocity of light in the polymeric medium (v) (Equation 2.2). The refractive index is reported at a specific wavelength, or frequency. For instance, it may be reported as n_D, which describes the refractive index of the polymer at 589 nm. The wavelength 589 nm is also known as the sodium D line, or the wavelength of yellow light from a sodium source.

$$n = \frac{c}{v} \tag{2.2}$$

A refractive index will only be provided for polymers transparent to light. Some common PV encapsulants are included in Table 2.3, along with their average refractive index. All these values are slightly larger than one. This indicates that the light slows down as it passes through the polymer. This decrease is due to scattering by the polymeric chains.

TABLE 2.3

Photovoltaic Encapsulant Materials and Corresponding Average Refractive Index across the Visible Spectrum

Polymers	Refractive Index
Polyethylene-*b*-polymethacrylic acid salt-*b*-polymethylacrylate	1.49
Polydimethylsiloxane (PDMS)	1.40
Ethylene vinyl acetate (EVA)	1.48

Sources: Data from J.E. Mark, 1999, *Polymer Data Handbook*, Oxford: Oxford University Press; J.E. Mark, 1996, *Physical Properties of Polymers Handbook*, Woodbury, NY: AIP Press.

There is an inverse relationship between refractive index and wavelength. This relationship is defined by Cauchy's formula, a polynomial function where A, B, C, and D are inherent material coefficients (Equation 2.3):

$$n = A + \frac{B}{\lambda^2} + \frac{C}{\lambda^4} + \frac{D}{\lambda^6} + \cdots \tag{2.3}$$

Unfortunately for PV manufacturers, the steepest changes for polymers occur over the visible light spectrum. However, these changes in refractive index are typically small. NuSil Silicones reports a 0.04 decrease in refractive index between 411 nm and 833 nm based on the following material coefficients: A = 1.522, B = 1.050e 4 nm², C = −9.789e8 nm⁴, and D = 1.312e14 nm⁶ for their LS-3354 silicone encapsulant [14]. Not all PV manufacturers report the wavelength dependence of refractive index on their technical data sheet. Therefore, a design engineer should consider these changes as part of their evaluation procedure.

A polymer's refractive index changes with temperature. There is no absolute correlation, but for most there is a slight decrease with increased temperature. These changes are typically small, on the order of $-10^{-4}/\mathrm{K}$ for polydimethylsiloxanes. An exception is some grades of polymethylmethacrylates (PMMAs) can become opaque at high temperatures.

The temperature dependence of the polymer's refractive index is typically not reported unless the polymer is sold for optical applications. Therefore, the absence of this data on the material data sheet should not be inferred as a lack of temperature dependence. A PV design engineer should perform refractive index measurements over the product's anticipated service life temperatures as part of their qualification testing.

Light is a diffuse medium that naturally luminates a surface at a variety of different angles. For simplicity, a single ray of light is depicted in Figure 2.6. Differences in refractive indices will change the light's trajectory when light

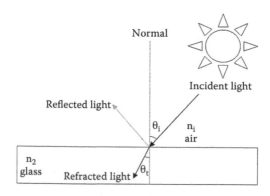

FIGURE 2.6
Depiction of the loss of light as a result of index mismatch.

travels between two media (e.g., glass and encapsulant). The angle of transmittance through the glass (θ_t) can be determined with Snell's Law requiring the refractive index of the air (n_1), the refractive index of the glass (n_2), and the angle of incidence (θ_i) of the incoming light (Equation 2.4):

$$n_1 \sin\theta_i = n_2 \sin\theta_t \tag{2.4}$$

In addition to refraction, there is a percentage of the incident light ray that is reflected from the substrate. The percentage of reflected light can be determined using the reflection coefficient, and it is commonly used to identify candidate materials for PV packaging. The reflection coefficient (R) is based on the principle that a large disparity in refractive indices will cause a larger reflection at the material interfaces. Therefore, reflection losses are estimated using the refractive index of the interface materials (n_1, n_2) and the angles of incidence (θ_i) and transmittance (θ_t) (Equation 2.5):

$$R = \frac{\left(n_1 \cos\theta_i - n_2 \cos\theta_t\right)^2}{\left(n_1 \cos\theta_i + n_2 \cos\theta_t\right)^2} \tag{2.5}$$

The percentage of transmitted light to the underlying cells is given by the transmittance coefficient (T), a dimensionless ratio (Equation 2.6):

$$T = 1 - R \tag{2.6}$$

Any two packaging materials will have slightly different refractive indices, and this mismatch will ultimately decrease the intensity of incident light on the underlying PV cells. Therefore, this relationship can help the design engineer optimize material selection or minimize processing errors by reducing reflection losses. As an example, 0.005% of normal incident light is reflected at an interface between glass and ethylene vinyl acetate (EVA) copolymer. Therefore, EVA is an excellent material selection for an optical coupling agent to glass. However, a 4% loss is expected if processing errors cause an air pocket between the glass and EVA.

An example calculation of perfect adhesion between glass and EVA under normal incidence ($\theta_i = 0°$):

Refractive index of glass (n_1): 1.50
Refractive index of EVA (n_2): 1.48

$$R = \frac{(1.48 - 1.50)^2}{(1.48 + 1.50)^2} = 5 \times 10^{-5} \times 100 = 0.005\%$$

An example calculation of imperfect adhesion between glass and EVA under normal incidence ($\theta_i = 0°$):

Refractive index of glass (n_1): 1.50
Refractive index of air (n_2): 1.00

$$R = \frac{(1.5 - 1.0)^2}{(1.5 + 1.0)^2} = 0.04 \times 100 = 4\%$$

PV manufacturers anticipate a 2% loss across all interfaces. Materials and processing must therefore be simultaneously optimized to reach this goal.

2.3.1.2 Yellowness Index

The color space is a graphical representation used to describe coloration. The two most common, listed in historical significance, are the Commission Internationale de l'Eclairage (CIE) XYZ color space developed in 1931 and redefined in 1976 and the L, a, b color scales developed by Hunter Laboratories in 1970. Both can be used to evaluate the yellowness index (YI). In this context, just as the name implies, the yellowness index is a quantification of the shade of yellow exhibited by a polymer. When yellowing is quantified, it will be reported either as ΔYI, referring to a CIE XYZ scale, or a ΔE shift, referring to a L, a, b scale.

This discussion will be limited to the yellowness index measured using ASTM D1925 based on the tristimulus (CIE) color scale [15]. This particular scale has been used extensively in the PV encapsulant literature. The XYZ system describes the blue, yellow-green, and red regions of the visible region of the electromagnetic spectrum. The visible portion of the electromagnetic spectrum is wrapped into a semicircle formation in the x–y plane (Figure 2.7) [16]. Red is described by a high x and low y, green is a high y and low x, and blue is a low y and low x value.

This ASTM specification requires an ultraviolet-visible (UV-Vis) spectrum be acquired. A UV-Vis spectrometer shines light upon the specimen at various wavelengths and monitors the intensity of the light transmitted through the polymer to a photodetector [17] (Figure 2.8). The percentage of light transmitted as a function of wavelength is called a UV-Vis spectrum (Figure 2.9).

The spectrum can be deconvoluted into the CIE color scale using tristimulus values (Figure 2.10). Tristimulus values are a mathematical description of the three primary colors, red, green, and blue, designated as X, Y, and Z, respectively (Equations 2.7 through 2.9). As an example, the unitless X_{CIE} tristimulus value is the transmission of the sample ($T(\lambda)$), the red component

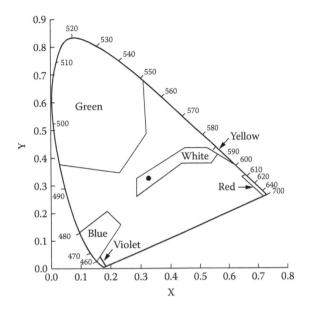

FIGURE 2.7
Commission Internationale de l'Eclairage (CIE) 1931 XYZ color space diagram in the *x–y* plane. (From W. Karwowski, 2001, *International Encyclopedia of Ergonomics and Human Factors*, Vol. 1. London: Taylor & Francis. With permission.)

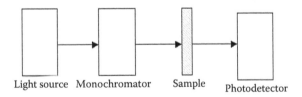

FIGURE 2.8
Ultraviolet-visible (UV-Vis) spectrometer components.

$(\bar{x}(\lambda))$, and Illuminant C $(S(\lambda))$ at each wavelength summed and multiplied by the chromaticity coefficient for the illuminant (K):

$$X_{CIE} = K \sum \bar{x}(\lambda) S(\lambda) T(\lambda) \tag{2.7}$$

$$Y_{CIE} = K \sum \bar{y}(\lambda) S(\lambda) T(\lambda) \tag{2.8}$$

$$Z_{CIE} = K \sum \bar{z}(\lambda) S(\lambda) T(\lambda) \tag{2.9}$$

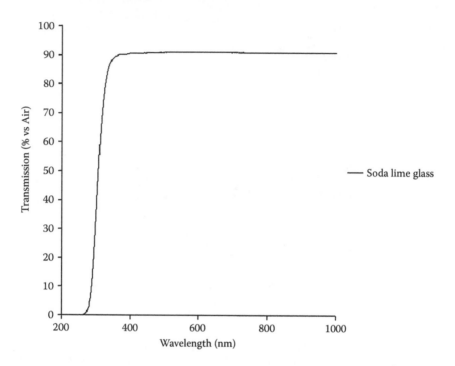

FIGURE 2.9
Typical ultraviolet-visible (UV-Vis) spectrum of soda lime glass in transmission mode.

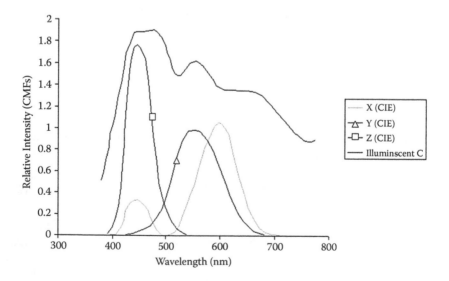

FIGURE 2.10
Tristimulus spectrum of the individual components and Illuminant C.

Once the CIE tristimulus values (X_{CIE}, Y_{CIE}, and Z_{CIE}) are calculated, the yellowness index can be determined (Equation 2.10). ASTM D1925 expresses the yellowness of the plastic relative to magnesium oxide. This comparison to magnesium oxide gives rise to the fixed coefficients (1.28 and 1.06) in the yellowness index equation:

$$YI = \frac{[100(1.28X_{CIE} - 1.06Z_{CIE})]}{Y_{CIE}} \tag{2.10}$$

In order for the plastic to appear yellowed, it must absorb light between 420 nm and 440 nm. A decrease in transmission will decrease the Y_{CIE}, increasing the yellowness index (YI). When subtracted from the original yellowness index (YI_o), the change in the index (ΔYI) increases (Equation 2.11):

$$\Delta YI = YI - YI_o \tag{2.11}$$

Researchers measure the change in color as a function of exposure time in order to gain insight into the chemical phenomena causing yellowing. Commonly, first-order degradation kinetics fit coloration changes caused by small organic molecules. Under first-order kinetics, the natural logarithm of the ratio of the YI graphed versus time (t), measured in hours (h), creates a slope defined by the rate constant (k_1), measured in inverse time, such as inverse hours (h^{-1}) (Equation 2.12). The rate constant can be used to predict the change in the yellowness index over time.

$$\ln(YI/YI_o) = -k_1 t \tag{2.12}$$

Color shifts in thermally aged EVA have been noted in a number of studies. Cuddihy and coworkers verified the color shift in Elvax® 150 was a first-order kinetic process described by a 10^{-4} h^{-1} rate constant. The polymer yellowed because of the small molecule peroxides, Lupersol 101, included in the commercial formulation. Under heat Lupersol 101, 2,5-dimethyl-2,5-*bis*(t-butyl peroxy) hexane, decomposes to form highly reactive peroxides. These peroxides create chromophores, making the EVA appear yellow [18]. Therefore, a kinetic analysis can be a useful failure analysis technique to understand the mechanism of polymeric yellowing.

2.3.2 Photovoltaic Module Performance

The polymer's optical properties directly influence the PV module's performance. However, it is not appropriate to just consider the optical properties of the individual materials in isolation. The total losses across the module's material stack determine the concentration of incident light on the embedded

PV cell. For this reason, the PV industry developed a series of functional tests for encapsulated PV cells to verify optical characteristics are optimized.

2.3.2.1 Quantum Efficiency Measurements

A quantum efficiency (QE) instrument measures the cell's efficiency for converting light into electricity. The QE instrument is composed of an amplifier, light source, filter wheel, chopper, monochromator, and ammeter. The cell is placed under a light source, and a monochromator shines defined wavelengths of light on the cell while the ammeter measures the produced current (Figure 2.11).

QE is the ratio of the total number of electrons (N_e) produced by the device versus the number of incident photons (N_v) at each wavelength of radiant light (Equation 2.13):

$$QE_\lambda = \frac{N_e}{N_v} \tag{2.13}$$

QE is an efficiency measurement defined by what is actually produced versus what is theoretically possible. Ideally, input would equal output ($N_e = N_v$), and there would be no wasted light, meaning the QE would have a value of 1. However, for various reasons, including the changes in refractive indices already discussed, the ideal curve does not occur (Figure 2.12).

The internal QE (IQE) curve describes the bare PV cell performance. Incident light with energy equivalent to the PV material's band gap forces electrons in the semiconductor material into their excited state. This creates

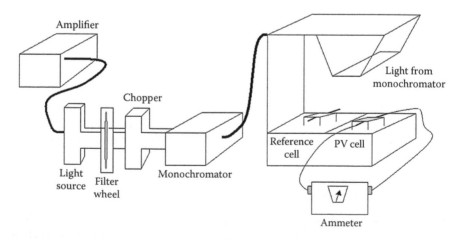

FIGURE 2.11
Components of a quantum efficiency (QE) instrument.

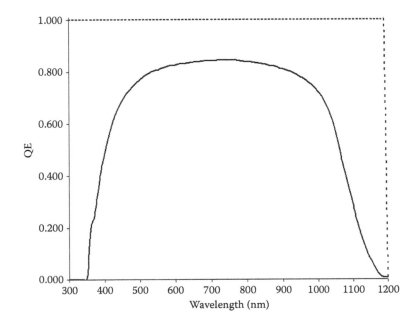

FIGURE 2.12
Ideal (dotted) and typical (solid) crystalline silicon quantum efficiency curve.

an electron-hole pair that must physically separate within the semiconductor. The electrons diffuse out of the semiconductor toward the electrical conduit. The flow of the electrons around the circuit creates a current. Once the electrons reunite with the holes, the reaction is complete. There are multiple causes for deviation from this ideal reaction. For instance, the electron-hole pair can immediately recombine inhibiting electron flow around the circuit. However, these inefficiencies in the IQE are rarely related to the polymer packaging.

The external QE (EQE) curve is the relevant diagnostic for packaging performance. The EQE is monitored on the back end of the manufacturing process. Instead of bare PV cells, the packaged cells are measured. The EQE describes deviations from ideal behavior as a result of transmission losses in the packaging.

Although the EQE curve of the stack is the optimal metric, device performance can be approximated with the IQE and the relevant UV-Vis transmission curves of the packaging components. Specifically, the changes in device performance can be estimated by multiplying the AM 1.5 irradiance at each wavelength by the transmission through the polymer at that wavelength and the device IQE at the specified wavelength (Equation 2.14):

$$Device\,Oulpul(\lambda) = Transmission(\lambda)\,Irr(\lambda)\,IQE(\lambda) \qquad (2.14)$$

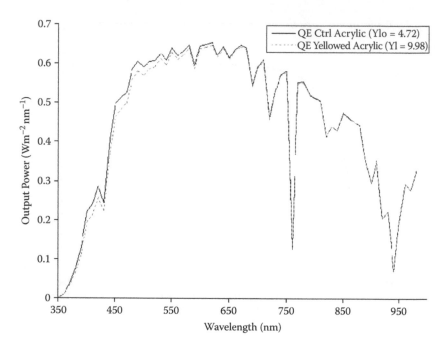

FIGURE 2.13
Example of estimated changes in power output of a crystalline silicon photovoltaic (PV) cell before and after acrylic yellowing (ΔYI = 5.26).

As an example, a five-unit change in yellowness index (ΔYI) for polymethylmethacrylate is estimated to cause a decrease of −0.02 to −0.04 watts per meter squared–nanometer (W/(m² • nm)) with the largest drop in the green-yellow region (Figure 2.13).

2.3.2.2 Current-Voltage (IV) Measurements

The current-voltage response of the packaged module is used to define the electrical characteristics reported on the product's data sheets. These measurements are performed in accordance with ASTM E1036 [19]. The standard defines experimental conditions as an irradiance of 1 kilowatt per square meter (kW/m²) of sunlight, 25°C (298 K) temperature, and wind speed of 1 meter per second (m/s). The irradiance should mimic the AM 1.5. This is a firm requirement for the region of the electromagnetic spectrum with the PV cell's highest QE response. If this light source is not accurately mimicked, the product will not meet consumers' expectations.

Under irradiance, the current is measured with zero applied voltage (I_{sc}) (Figure 2.14). The current is then monitored with an increasing voltage until there is no current flow (V_{oc}). This maps out the current-voltage (IV) response of the packaged module.

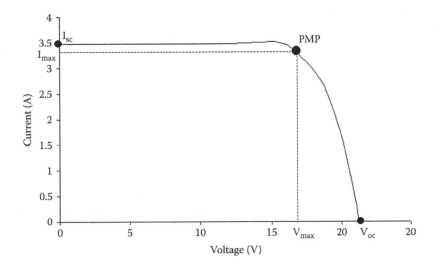

FIGURE 2.14
Typical current-voltage (IV) curve.

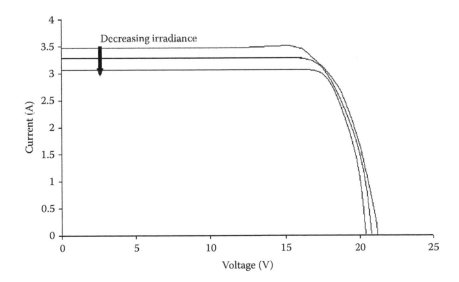

FIGURE 2.15
Expected trend in the current-voltage (IV) curve with decreasing light irradiance.

The current from the module changes proportionally with the irradiance of the incident light [20,21]. Visible light irradiance can decrease due to transmission loss at the material interfaces. For instance, any changes along the edges of the encapsulation, such as yellowing or peeling, will lower the device current by decreasing the solar irradiance on the PV cells (Figure 2.15).

Open circuit voltage rarely changes due to packaging considerations. The two exceptions are small molecule migration from polymeric encapsulants and high transmission of infrared radiation. The open circuit voltage is predetermined by the semiconductor's band gap. Changes in the semiconductor formulation are the dominant cause of voltage drift (Figure 2.16). At high temperatures, small molecules can migrate out of the polymeric substrates. These migrants can potentially react with the semiconductor materials, changing the open circuit voltage. This phenomenon has been cited as a cause for electrical drift when there were no signs of encapsulant delamination or yellowing [22]. However, research in this area is sparse. Infrared light transmission can increase device temperature and decrease voltage. Different PV cells have different thermal responses. However, a 5% to 7% decrease in voltage with every 10 K increase in temperature is typical. Therefore, ideal packaging materials would decrease infrared radiation without interfering with visible light transmission.

The instantaneous power (*P*), measured in watts (W), is the product of current (*I*), measured in amperes (A), at a specified voltage (*V*), measured in volts (V), at a specific time (Equation 2.15):

$$P = VI \qquad (2.15)$$

The peak maximum power (PMP, P_{max}) point occurs at the curve maximum, where the product of current (I_{max}) and voltage (V_{max}) is highest. Ideally, to maximize power production, the device would simultaneously operate at the open circuit voltage (V_{oc}) and short circuit current (I_{sc}). Because this is

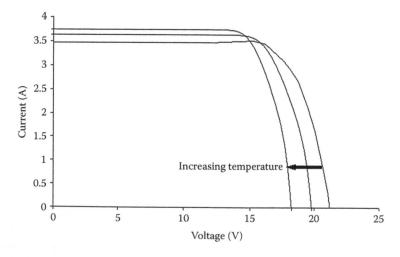

FIGURE 2.16
Expected trend in the current-voltage (IV) curve with increasing temperature.

a physical impossibility, the best the device can produce is PMP. The ratio between actual and ideal behavior is the fill factor (*FF*) (Equation 2.16):

$$FF = \frac{P_{max}}{P_{ideal}} = \frac{I_{max}V_{max}}{I_{sc}V_{oc}}$$

(2.16)

A typical fill factor for commercial products is 0.6 to 0.8 for inorganic solar cell chemistries.

PV manufacturers use the packaging factor to verify packaging integrity for each module. The current-voltage characteristics are measured for the bare PV cells produced on the front end of the process. They are remeasured on the packaged module at the end of the manufacturing process. The peak maximum power of the bare substrate (*PMP*$_{BE}$) subtracted by the peak maximum power of the packaged module (*PMP*$_{FE}$) divided by the bare substrate PMP is called the packaging factor (*PF*) (Equation 2.17):

$$PF = \frac{PMP_{BE} - PMP_{FE}}{PMP_{FE}}$$

(2.17)

Successful material selection and processing conditions are denoted with a positive packaging factor.

The module efficiency is an important consideration when a consumer makes a purchase. High-efficiency modules have a higher market appeal for consumers in geographical regions with low solar irradiance, such as the northern United States. A higher efficiency allows a consumer to obtain more power at lower irradiance and fewer hours of sunlight. They also have general market appeal because a consumer can purchase fewer modules and obtain more power from their array. The efficiency is a percentage measurement of the power produced ($P_{max} = I_{max}V_{max}$) divided by the total power of incident sunlight on the module (P_{irr}) (Equation 2.18):

$$\eta = \left(\frac{I_{max}V_{max}}{P_{irr}} \right) 100\%$$

(2.18)

The total power of incident sunlight is equivalent to the active semiconductor area, in square meters (m²), multiplied by the irradiance, 1 kW/m² for ASTM E1036 (Equation 2.19):

$$P_{irr} = (Area)(1 kW/m^2)$$

(2.19)

The current industrial standard for module efficiency is 12% to 18% efficiency. However, this value varies significantly based on the encapsulated PV cell chemistry.

2.4 Thermal Properties

Heat is a form of thermal energy caused by molecular motion. When applied to polymers, heat increases polymer chain motion and causes the polymers to undergo morphological changes manifesting in dimensional, mechanical, and electrical property shifts. Engineers must compensate for these changes in their design by estimating the magnitude of the property shift from material characterization techniques. Functional tests are used in the solar industry to verify desired performance based on these estimations.

2.4.1 Material Properties

Heat can be absorbed or conducted through a polymer sample. The various classifications of polymers have different thermal characteristics. For instance, ionomers can exhibit three transitions, while amorphous polymers only exhibit one. This section will expand on the discussion from Chapter 1 by discussing the methodology used to identify material transitions. In addition, characterization of polymeric heat conduction will be discussed.

2.4.1.1 Morphology

Morphology describes the three-dimensional organization of polymer chains. The three morphological classifications relevant for this discussion are semicrystalline, amorphous, and ionic clusters. This behavior is characterized in polymer science with the use of differential scanning calorimetry (DSC). DSC requires a small sample size, on the order of a few milligrams (mg), placed in an aluminum sample pan (Figure 2.17). The sample pan is placed in a controlled chamber and heated at a user-specified ramp rate, recorded in Kelvin per minute (K/min). During the ramp, the heat flow inside the chamber is monitored. As the polymer sample undergoes morphological changes, it will absorb or emit heat based on whether the phase transition is endothermic or exothermic, respectively. This heat absorption or emission is captured as a change in heat flow, measured in watts per gram of material (W/g). For instance, the polymer absorbs heat when the crystalline domains melt; therefore, melting events are identified as a drop in the instrument heat flow (Figure 2.18). Conversely, when the polymer recrystallizes, there is a release of heat from the sample denoted with a rise in the chamber's heat flow. Both transitions occur over a temperature range. The apex of the curve is typically reported as the transition temperature. Integration of the area under the transition defines the heat capacity [23]. In order to be a valid first-order transition, it must be reversible. Therefore, melt and crystallization temperatures can be observed on multiple heating and cooling ramps. For this reason, most DSC analyses are performed with at least two heating cycles.

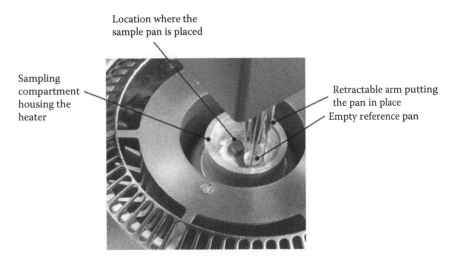

FIGURE 2.17
Image of the TA Q Series™ Instrument. (Figure courtesy of TA Instruments-Waters LLC.)

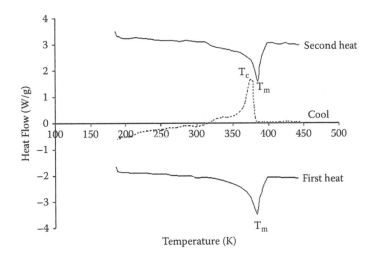

FIGURE 2.18
A typical differential scanning calorimetry (DSC) curve for a semicrystalline polymer.

The second-order, glass transition (T_g), is also visible in a DSC curve. As an inherent polymeric property, it is a reversible transition identified on multiple thermal cycles as a slight change in the slope of the heat flow curve (Figure 2.19). The temperature at the point where the slope changes defines the glass transition. The absence of a glass transition in the DSC spectrum does not mean the polymer does not have one. A number of glass transitions

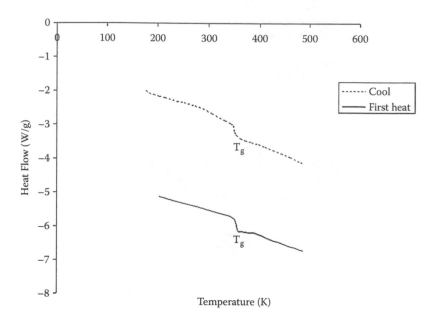

FIGURE 2.19
A typical differential scanning calorimetry (DSC) curve exhibiting a T_g.

are subambient; therefore, a wide temperature range must be swept to insure all transitions are identified.

An irreversible transition is likely a molded in stress, which developed during processing. It is common for both amorphous and semicrystalline thermoplastics to exhibit molded-in stress. This processing flaw is discussed in detail in Chapter 4. Briefly, when the polymer flows into the mold, the chains are elongated in the melt flow direction. The elongated chains solidify in small, oriented domains. When the specimen is heated during a DSC cycle, the polymer chains relax from this oriented configuration. Because the same flow geometry is not induced in the DSC pan, this transition is not reversible and is not an inherent characteristic of the polymer resin. Therefore, the relaxation will only be present in one heating cycle as an exothermic peak adjacent to the glass transition, and it will not be reproducible in subsequent cycles of the same sample [24].

2.4.1.2 Coefficient of Thermal Expansion

The glass transition is also called the softening point of the material. Almost all polymers have a glass transition (T_g). Above T_g, the morphology of the polymer changes as the polymer chains slide past one another and find a new equilibrium position. Due to this morphological change, material properties are different on each side of the T_g. Therefore, changes to the polymeric properties must be anticipated if the service temperature of a product spans the glass transition temperature.

The coefficient of thermal expansion (CTE) has different values above and below the glass transition. The CTE (α) is the partial differential in volume (V) as a function of temperature (T) at constant pressure multiplied by inverse volume (Equation 2.20). It is reported as inverse temperature, typically inverse Kelvin (K^{-1}).

$$\alpha = \frac{1}{V}\left(\frac{\partial V}{\partial T}\right)_P \tag{2.20}$$

Glass transition temperatures and coefficients of thermal expansions will rarely be reported on product data sheets from suppliers. They are inherent polymer properties independently measured by the end user or approximated from the values reported in the tables of industrial handbooks. Some relevant data for polymers used in solar packaging are included in Table 2.4 [13]. These values are highly dependent on the polymer's formulation, and they should be used as estimates. An empirical measurement of the relevant polymer grade should be performed when possible.

The thermal expansion of most materials can be measured with dilatometers. A dilatometer records changes in volume due to increased temperature. The two most common measurements are capacitance and strain. In both experiments, samples are placed between parallel plates. With a capacity dilatometer, the capacitance between the plates decreases as the specimen expands with increasing temperature. Alternatively, for a connecting rod dilatometer, a strain gauge connects the two parallel plates. The specimen expands with increasing temperature to cause an increase in measured

TABLE 2.4

Polymer Classifications, Names, and Corresponding Thermal Expansion Coefficient

Classification	Polymer Name	Thermal Expansion Coefficient (K^{-1})
Thermoplastics	Polyethylene terephthalate	9.1×10^{-5}
	Ethylene vinyl acetate copolymer	$25–16 \times 10^{-5}$
	Polyvinyl fluoride	9×10^{-5}
	Polyethylene-*b*-polymethacrylic acid salt-*b*-polymethylacrylate	$5.9–5.7 \times 10^{-5}$
Elastomer	Polydimethylsiloxane	$7.9–5.9 \times 10^{-4}$

Source: Data from J.E. Mark, 1996, *Physical Properties of Polymers Handbook*, Oxford: Oxford University Press.

strain. Detailed procedures for the measurement of the thermal coefficient of expansion via dilatometry can be found in ASTM D696 [25].

Thermomechanical analysis (TMA) is a connecting rod dilatometer commonly used to measure polymer thermal expansion coefficients. Both the glass transition temperature and the coefficient of thermal expansion can be measured using TMA. During the experiment, the expansion probe is placed on the surface of the polymer with a nominal force (~1000 micronewtons [μN]) applied to the largest surface area (Figure 2.20). The polymer temperature is ramped in an inert atmosphere over the temperature range of interest. The increased temperature causes the polymer to expand, resulting in a vertical displacement of the probe. The coefficient of thermal expansion is calculated from the initial volume and the changes in displacement divided by the change in temperature, also known as the slope of the TMA curve (Figure 2.21). A discontinuity in the curve is present at the T_g of the polymer. The thermal expansion coefficient is measured on each side of the T_g.

2.4.1.3 Thermal Conductivity

Thermal conductivity is a quantitative measurement of the material's ability to remove heat from an adjacent heat source. The empirical measurement is most commonly done under static conditions. A static experiment requires thermal conductivity be measured after the test substrate has reached an equilibrium temperature with the heat source.

Those familiar with polymer packaging in the electronic industry are accustomed to the measurement of the *apparent thermal conductivity*. This is a static technique that requires deriving the thermal conductivity from the material's thermal resistance (R_{TH}), measured in Kelvin per watts (K/W). Per ASTM D5470 [26], the thermal resistance is defined as the temperature difference (ΔT), measured in Kelvin, across the polymer substrate divided by the dissipated thermal power (Q), measured in watts (W) (Equation 2.21):

$$R_{TH} = \frac{\Delta T}{Q} \tag{2.21}$$

The polymer's ability to remove heat, thermal conductivity (α_{TH}), measured in watts per meter–Kelvin (W/(m • K)), can be evaluated using the thermal resistance and the polymer's geometric dimensions. It is the thickness of the polymer (d), measured in meters, divided by the product of the thermal resistance and the polymer's area (A), measured in square meters (m²) (Equation 2.22):

$$\alpha_{TH} = \frac{d}{R_{TH} A} \tag{2.22}$$

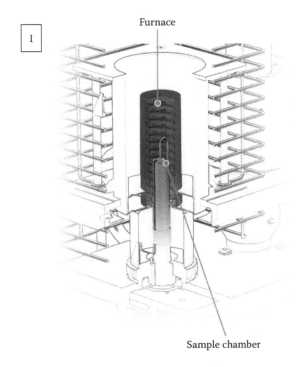

I

Furnace

Sample chamber

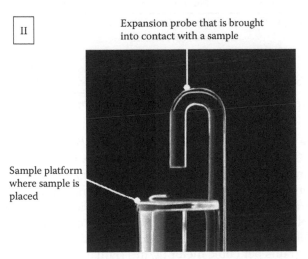

II

Expansion probe that is brought into contact with a sample

Sample platform where sample is placed

FIGURE 2.20
(I) Depiction of a thermomechanical analysis (TMA) sample compartment. (II) Image of the TA Q Series™ instrument. (Figure courtesy of TA Instruments-Waters LLC.)

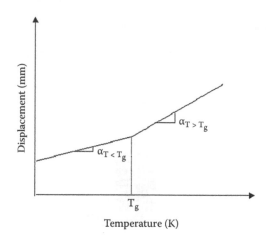

FIGURE 2.21
Typical thermomechanical analysis (TMA) curve used to measure the polymeric thermal expansion coefficient.

TABLE 2.5

Polymer Classification and Thermal Conductivity of Representative Polymers

Classification	Polymer Name	Thermal Conductivity W (m • K)$^{-1}$
Thermoplastics	Polyethylene terephthalate	0.15
	Ethylene vinyl acetate copolymer	0.34
	Polyvinyl fluoride	0.14
Elastomer	Polydimethylsiloxane	0.25

Source: Data from J.E. Mark, 1996, *Physical Properties of Polymers Handbook*, Woodbury, NY: AIP Press.

DSC can be used to measure the thermal conductivity of a polymer sample when the DSC is operated in modulation mode. A modulated heat flow is a cyclic heat flow, described by the frequency (ω = 2π/period (P)), measured in units of inverse seconds (s^{-1}), over a specified temperature range. Thermal conductivity (α_{TH}) can be calculated from specific heat (C_p), measured in joules per Kelvin–gram (J/(K • g)); apparent heat capacity (C), measured in joules per Kelvin (J/K); the sample's area (A); and the sample's density (ρ), measured in grams per cubic meter (g/m^3) (Equation 2.23):

$$\alpha_{TH} = \frac{2\pi C^2}{\rho A^2 P C_p} \tag{2.23}$$

Most polymers used in solar applications have similar thermal conductivities (Table 2.5) [13]. As will be described further in Chapter 3, the inherent polymer thermal conductivity can be altered by the addition of various additives compounded into the formulation.

2.4.2 Photovoltaic Module Performance

PV modules are characterized by changes in electrical performance due to changes in ambient temperature. A product data sheet for PV modules includes a thermal properties section containing the temperature dependence of short circuit current (I_{sc}), open circuit voltage (V_{oc}), and peak maximum power (PMP). These assertions are based on the temperature behavior of the semiconductor material used in the solar cells. However, the plastic packaging can also undergo changes during cyclic service temperatures.

2.4.2.1 Frame Warp

Like most materials, polymers expand when heated. During material selection, it is imperative to consider the materials adjacent to the polymer. Combining the inherent thermal and mechanical properties, it is possible to identify the highest contributor to the thermal expansion in the assembly. The contribution of each material can be estimated by multiplying the elastic modulus (E) by the sample thickness (t) and the coefficient of thermal expansion (α) (Equation 2.24). A dimensional tolerance of the design will conclude if there is enough space to allow for a polymer's expansion during thermal cycling.

$$\sim \alpha t E \qquad (2.24)$$

For thermal processes, it is important to consider the service temperature range. If the polymeric components undergo a phase transition during the service temperature, the CTE on both sides of the transition must be included in a tolerance analysis.

Frame warping is commonly the result of a miscalculation during tolerance analysis. Factors such as unforeseen environmental contaminates, swelling of polymeric substrates, and processing-induced instability can all contribute to the introduction of unexpected stresses on the frame.

Polymers are susceptible to swelling from various organic molecules present in nature and adjacent substrates in the assembly. Permeant diffusion into the substrate can increase the polymer thickness beyond the intended design. This unplanned dimensional change is rarely included in the initial tolerance analysis and can create field failures. It can be exposed as a potential issue by evaluating the chemical compatibility of the adjacent substrates. This is typically performed by placing those materials together and thermal cycling the assembly across the service temperature range. Changes in color, mechanical strength, or dimensions are typically deemed an incompatibility. Incompatible materials are generally those with similar chemical compositions. In this instance, there will be a free exchange of molecules across their interface due to the similarity in their chemical structure.

Thermoplastics can exhibit shrinkage depending on the processing conditions used during molding. This possibility is best evaluated by performing a dimensional analysis before and after exposing the molded part to thermal cycling.

Similarly, a common failure mode during thermal cycling is cracking and delamination caused by thermal mismatch between the polymer and the adjacent substrates. Both failures result in a decrease in module power. A crack in the solar cell results in an immediate failure for that electrical string. A delamination creates an air gap and a loss of collected light due to reflectance at the interface.

2.4.2.2 In-Service Temperature Measurements

Critical polymer components included in solar module assembly typically have a certified relative temperature index (RTI) rating based on their performance on tests included in UL 746B [27]. The measurement is a thermal endurance technique used to estimate the polymer's lifetime at elevated temperatures. The RTI is the maximum service temperature at which a critical material property is not compromised due to thermal degradation. There are three RTI classifications based on material property: electrical, mechanical strength, and mechanical impact. At timed intervals, the polymer is removed from thermal chambers, and properties are evaluated for changes. The critical property must not degrade by more than 50% of its initial value during the test. Because this is an endurance measurement, it is one of the most time-intensive UL specifications, typically requiring 7 to 12 months to complete. For this reason, it is important to choose precertified polymers if there is a short development cycle.

Based on IEC 61730 [28,29] and UL 1703 [30], encapsulants, enclosures, and supports for live parts must have a minimum electrical and mechanical RTI of 20 K above the maximum operating temperature. Polymers included as superstrates and substrates (e.g., backsheets) must have an RTI equal to or greater than 90°C (363 K), or 20 K above the operating temperature, whichever is higher [31].

2.5 Mechanical Properties

Due to short development timelines, PV manufacturers must often choose polymeric candidates based on information provided on the manufacturer's data sheet. However, these values are obtained under controlled laboratory conditions. Therefore, PV design engineers should be aware of any experimental simplifications that could cause a miscalculation of in-field product performance.

2.5.1 Material Properties

The polymeric mechanical properties reported on the technical data sheets may or may not be what is required for the material evaluation of an application. Ultimately, it is the responsibility of the design engineer to identify the mechanical properties required for module packaging. The easiest method for determining these criteria is a combination of computer simulation and empirical testing.

When choosing materials, design engineers must specify the critical performance property for their application. A finite element analysis (FEA) is a computer program that solves a series of mechanical equations to help identify the mechanical deformation required for failure. Compressive, tensile, and impact stress are the most common deformation modes for components included in PV modules. Once a mode is identified, the FEA will simulate the numerical threshold required to induce failure. As an example, FEA may conclude a compressive snow load greater than 4000 pascal (Pa) will cause frame buckling. This is called the failure requirement. This threshold in mechanical properties would be used to select a material for further evaluation. For instance, any polymer with inherent properties below the failure requirement would not be a logical choice for further evaluation.

Once the materials have been narrowed based on their inherent properties, the list can be further reduced based on the environmental endurance of the critical performance properties. A number of environmental stresses (i.e., temperature, wind, humidity, etc.) work simultaneously and synergistically to chemically age polymeric components. The deterioration of mechanical performance due to environmental exposure cannot be predicted with an FEA. How all these stresses combine to change performance is best empirically measured on actual parts in a controlled laboratory environment. After environmental conditioning, mechanical experiments are repeated on test specimens to quantify the drop in mechanical performance. The expected material degradation helps the design engineer specify the design requirement. The design requirement is the additional performance above the failure requirement necessary to compensate for property deterioration during service.

Ultimately, materials are chosen based on their margin of safety, the design requirement divided by the failure requirement minus one (Equation 2.25):

$$\text{Margin of Safety} = [\text{Design Requirement}/\text{Failure Requirement}] - 1 \quad (2.25)$$

The margin of safety is the additional mechanical performance desired over that required. Most PV manufacturers require a minimum margin of safety of one for polymers used as structural components.

In order to select polymers for qualification testing, the design engineer must be familiar with the mechanical tests performed by polymer manufacturers and reported on data sheets.

2.5.1.1 Durometer

A durometer is a quantitative measurement of a polymer's hardness. It is a commonly used technique in a number of industries for design and quality control. Its popularity stems from quick analysis and ease of use. A measurement can be completed in a few seconds with little technical training.

A polymer's hardness is measured using the Shore hardness scale. There are multiple Shore scales designated as A through F. Different scales are appropriate for measuring different polymer classes. Shore A and D scales are relevant for most polymers included in PV module packaging. Shore A is used for soft polymeric materials and elastomers; and Shore D is used for harder polymers, thermoplastics, and thermosets. This discussion will be limited to this range.

The spring-loaded instrument used to make the measurement is appropriately referred to as a durometer. There are a number of manufacturers of durometers, including Instron (Norwood, Massachusetts) and Rex Gauge Company (Buffalo Grove, Illinois). All durometers are composed of three principal parts: a spring, a sensing pin, and a digital or dial display, ranging from 0 to 100 (Figure 2.22).

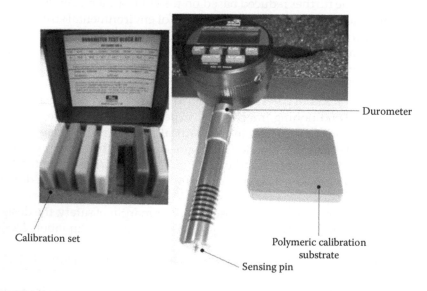

FIGURE 2.22
Image of Shore A durometer and calibration substrates.

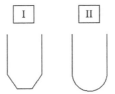

FIGURE 2.23
Depiction of the sensing pins used for thermoplastics, Shore D (I), and elastomers, Shore A (II).

The sensing pin and spring are dependent on the Shore range required. A Shore A durometer will have a low-force constant spring and a flat-cone sensing pin to minimize substrate penetration. In contrast, Shore D durometers have a large-force constant spring and a sensing pin with a sharp cone point (Figure 2.23). A series of test substrates with known durometers are typically purchased with the instrument for calibration.

Both scales are included in ASTM D2240 [32]. The test specimen, at least 6 millimeters (mm) thick, is placed on a flat surface, and the sensing pin is pressed into the polymer. The specimen's resistance to deformation will cause compression of the spring resulting in a reading between 0 and 100. Based on the ASTM standard, only numbers between 10 and 90 are valid. Higher readings correspond to harder test specimens.

Durometer measurements are commonly used in quality control of adhesives and thermosets. The chemistry of these classifications requires two components to react in a precise ratio to form a cross-linked structure. Any deviation from the specified mix ratio will cause incomplete cure. The manufacturer's specified cure time is the length of time required for the chemical reaction to occur and the final material properties to manifest. The durometer readings must level off to the manufacturer's specification by the end of the cure time in order to verify proper cure. Therefore, the durometer is a quality check to verify the material has been appropriately dispensed during processing.

2.5.1.2 Peel Strength

Peel strength is the most common mechanical procedure to evaluate bond strength between two substrates. Details for experimental conditions, including the strain rate, sample preparation, and conditioning are available in ASTM D903 [33]. The method of analysis is referred to by the direction of applied force to induce peel. The most common configurations are 180-degree and 90-degree peels (Figure 2.24). Peel strength is the average force required to cause failure divided by the width of the bond and reported in force per unit width, such as newtons per meter (N/m). Product data sheets will include the measured value, the experimental temperature, and the substrates used to evaluate peel strength.

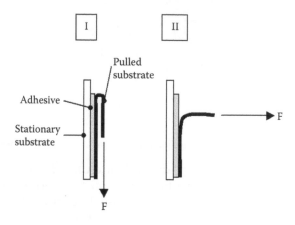

FIGURE 2.24
Sample orientations for (I) 180° and (II) 90° peel testing.

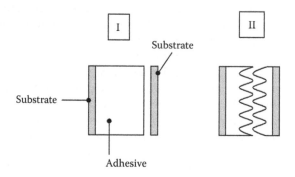

FIGURE 2.25
Depiction of (I) adhesive and (II) cohesive failures.

In addition to these experimental variables, the failure mode is commonly reported. Adhesive and cohesive failures are the two types of failure observed during peel testing (Figure 2.25). Adhesive failure indicates the bonding interface between the adhesive and the substrate as failed. Cohesive failure indicates internal failure of the adhesive or, more rarely, the substrate. Often there is not a single mode of observed failure. In these instances, the percentage area of adhesive or cohesive failure will be reported.

2.5.1.3 Tensile and Compression

Instron, Shimadzu (Columbia, Maryland), and Imada (Northbrook, Illinois) all make instruments for testing polymeric mechanical behavior. These instruments are composed of a motorized stage coupled with displacement and force gauges (Figure 2.26). As defined in Chapter 1, displacement is

FIGURE 2.26
The experimental setup used in tensile testing. (Figure courtesy of Instron® Instruments.)

converted into engineering strain by dividing by the initial sample length. Force is converted to engineering stress when divided by the initial area.

Tensile behavior is the more commonly reported mechanical behavior in technical data sheets. It is a necessary characterization when the application requires the polymer to be simultaneously pulled in opposite directions. In order to comply with ASTM D638 [34] for tensile testing, it is necessary to cut test specimens into specific geometrical shapes, termed *dog-bones* (Figure 2.27). The jaws of the mechanical grips hold the flattened areas of the shape so that the thinner cross section is the only area under elongation.

The sensitivity of the experiment is dependent on the load cell used in the frame. As previously mentioned, thermosets, thermoplastics, and ionomers require a high amount of force to elongate. Therefore, a large load force is commonly used, 5 to 10 kilonewtons (kN). In contrast, elastomers require less force for the same amount of elongation. These are commonly tested with a lower load cell, such as 1 to 2 kN.

Compression measurements are performed by squeezing the polymer between two plates. The same apparatus is used for compression and tension; however, the direction of applied force is in opposite directions (Figure 2.28). ASTM D695 [35] should be consulted for the precise geometric dimensions for this test. Like tensile measurements, a higher load force will be required for thermosets, ionomers, and thermoplastics, and a lower load force is appropriate for elastomers. Unlike tensile tests, a compression curve does not exhibit a break point (Figure 2.29). When the specimen has flattened, the stress increases exponentially. During this exponential rise, the plates begin to press against each other because the sample's resistance has been overcome.

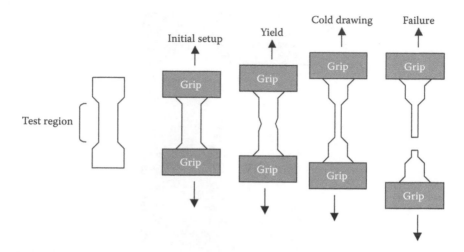

FIGURE 2.27
Example of the dimensional changes in a dog bone–shaped thermoplastic during tensile testing.

FIGURE 2.28
The experimental setup used in compression testing. (Figure courtesy of Instron Instruments.)

The tensile and compression curves described above are dynamic measurements. However, they both can be performed statically. There are two tests used for static mechanical evaluation: stress relaxation and creep. Both are common mechanical behavior techniques for polymers, but these properties rarely appear on the data sheets from polymer manufacturers.

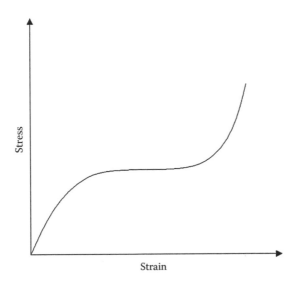

FIGURE 2.29
Example of a typical compression curve.

During stress relaxation, the specimen is held at a constant strain, and the change in stress is monitored as a function of time (Figure 2.30). The stress will decrease as the polymer chains rearrange to a new equilibrium in the specimen. This property is not commonly used in PV applications.

Creep testing is performed by applying a constant stress and then measuring the polymer's strain over time (Figure 2.31). It is a priority design consideration for polymers placed under a static load (e.g., weight). Details for test parameters can be found in ASTM D2990 [36]. A typical data collection is 100 to 1000 hours with extrapolation to the desired service life. The percent change in strain and experimental parameters, such as stress and temperature, are reported in the experimental results.

Both stress and temperature can affect the observed outcome of creep testing. The initial stress must be lower than the ultimate yield stress to be specified as creep. The higher the stress, the larger is the expected change in strain. Similarly, the higher the specimen temperature, the larger is the change in strain. Care must be taken to not perform a test at the polymer's transition temperature, unless the transition is transversed during the product's normal service. Therefore, measurements are typically performed 10 to 20 K away from the transition in order to minimize its effect.

When constant stress measurements are performed on elastomers, they are referred to as compression sets. ASTM Standard D395 [37] outlines the experimental protocol for these measurements. Typically, the elastomer is compressed to 25% of its original height under a controlled temperature (296, 343, 394, or 423 K) for a specified duration (22 h, 70 h, 168 h, or 1000 h).

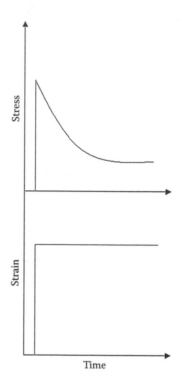

FIGURE 2.30
Example of a stress relaxation curve.

Usually, elastomers return to their initial dimensions after the stress is removed. The polymer will lose restorative properties if the peak stress is high enough to break chemical bonds. Compression set is a measurement of the permanent deformation of the elastomer after the applied stress is removed. It is reported as the percentage of the height not restored once the weight is removed. For instance, a 100% compression set indicates there is no measurable recovery in the elastomer to its initial height.

2.5.1.4 Impact Resistance

Polymer manufacturers rarely specify impact resistance. Polymers used in barrier applications (e.g., windshields and safety glasses) are the exception. When reported, impact testing is commonly performed in accordance with ASTM D256 [38]. Izod impact testing requires a rectangular specimen with a notch cut out of the center. A swinging pendulum strikes the specimen at the position of the notch while it is held vertically above the jaws of a grip (Figure 2.32). The energy absorbed when the sample breaks reduces the pendulum's momentum. The changes in momentum allow the engineer to calculate the impact resistance, measured in units of joules per meter (J/m).

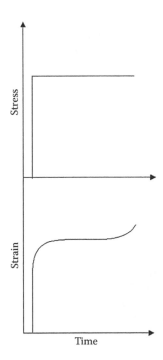

FIGURE 2.31
Example of a creep curve.

2.5.1.5 Flexural Testing

Arguably, flexural testing is the most relevant mechanical measurement for PV applications. The test was designed to measure the polymer's resistance to increasing load. ASTM D790 [39] specifies a test specimen of uniform dimensions supported across two parallel beams separated by a given length, termed the *support span*. The experimental setup is defined by the number of inflection points along the support span (Figure 2.33). It is standard to perform a three-point bend where the force is applied in the center of the span. Polymers that fail to adequately deform under this configuration are tested under a four-point test method, with two inflection points equally spaced between the support beams. In both configurations, a force is applied at a constant rate causing the top surface to experience a compressive load and the bottom test surface to experience tension. The applied force is monitored as a function of displacement until 5% deflection is reached or the specimen breaks.

The equations for flexure modulus and strength are dependent on the experimental setup and sample geometry. A list of relevant equations is included in ASTM D790 [39]. Using a three-point bend as an example, the flexural modulus is reported in units of pascal (Pa = 1 N/m²). For a rectangular

FIGURE 2.32
The experimental setup for Izod impact testing. (Figure courtesy of Instron Instruments.)

specimen, it can be calculated from the slope (n) of the load versus deflection curve, measured in newtons (N/m), and the specimen width (b), its thickness (h), and the support span (S), all measured in meters (m) (Equation 2.26):

$$E = \frac{S^3 n}{4bh^3} \tag{2.26}$$

The flexural strength is the highest stress the sample experiences prior to breakage. The strength is reported in pascals and is calculated from the peak force (P), measured in newtons, and the geometric factors, including support span, width, and thickness, measured in meters (Equation 2.27):

$$\sigma = \frac{3PS}{2bh^2} \tag{2.27}$$

2.5.2 Photovoltaic Module Performance

Polymer manufacturers report mechanical behavior under controlled temperature and humidity. These controlled test conditions do not allow design

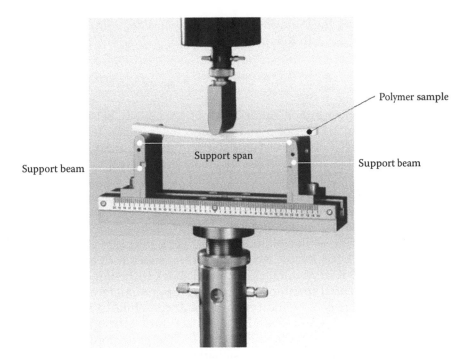

FIGURE 2.33
Depiction of a three-point flexural test. (Figure courtesy of Instron Instruments.)

engineers to predict how the design requirement will change under environmental exposure. After installation, a PV module will be exposed to various chemical and mechanical stresses. Acid rain, soil, and sea mist are common chemical stimuli, and hail, wind, and temperature extremes are common physical stimuli. Therefore, the PV industry must perform a number of practical tests to verify the safety factor is not compromised under weathering.

2.5.2.1 Thermal Cycling and Humidity Testing

A practical consideration when designing experiments is the geographical region with the largest product deployment. Service temperature differentials are highly dependent on the geography. Tropical climates are subject to the largest temperature extremes. It is not uncommon to find locations with temperature differentials as large as 35 to 55 K during the same day.

For flat module displays, the module temperature is typically 10 to 20 K above ambient temperature at peak sunlight. The module temperature (T_{module}), in Kelvin (K), is a function of solar irradiance (I_{irr}), in watts per square meter (W/m^2), the wind speed, in meters per second (m/s), and ambient temperature ($T_{ambient}$), in Kelvin (Equation 2.28):

$$T_{module} = I_{irr}(e^{(a+b(windspeed))}) + T_{ambient} \qquad (2.28)$$

The constants are determined empirically but are usually between –3.58 to –2.98 and –0.130 to –0.0455 for a and b, respectively, when temperature is measured in degrees Celsius (°C) [40–42]. It is typical for a module to spend at least one-third of its lifetime at elevated temperatures [43].

These practical considerations are not reflected in most testing standards. For instance, ASTM E1171 [44] has been adopted by the PV industry as a standard for conditioning samples prior to performing mechanical evaluations. This standard specifies thermal cycling (–40 to 85°C/233 to 358 K), humidity-freeze (85% relative humidity [RH], –40°C/85% RH, 233 K), damp heat (85°C/85% RH, 358 K/85% RH), and thermal shock (–40 to 110°C in 20 minutes/233 to 383 K in 20 minutes). Thermal shock is the largest proposed temperature differential spanning 150 K. This is approximately three to four times the most severe geographical environment. Therefore, to avoid overengineering the packaging materials, PV manufacturers will supplement this testing with the thermal gradient expected for the geographical locations with the highest number of module installations.

The ASTM standards specify environmental conditioning; however, there is limited direction on evaluating product performance. Customers expect electrical output to meet the technical specifications listed on the module's data sheet. Based on a survey of several module manufacturers, peak power is expected to drop less than 10% during service (Table 2.6) [45–47]. To verify compliance with this expectation, PV manufacturers condition their modules according to these ASTM standards. Prior to and after conditioning, the module's packaging is examined to verify there are no cracks or delamination that would compromise the module's electrical performance.

TABLE 2.6

Photovoltaic Companies, Selected Modules, and Peak Power Warranties

Company and Module ID	Peak Power and Tolerance
SunPower 225 Solar Module	225 W ± 5%
BP Solar SX 3200	200 W ± 9%
First Solar FS-270	70 W ± 5% for the first 10 years

Sources: Data from SunPower, 2008, Technical Data Sheet for 225 Solar Module (http://www.sunpowercorp.com); BP Solar, 2007, Technical Data Sheet for SX 3200 (http://www.bpsolar.com); First Solar, 2009, Technical Data Sheet for First Solar FS Series 2 PV Module (http://www.first solar.com).

2.5.2.2 Salt Fog

Similar to thermal cycling and humidity soak, standards for salt fog are used to condition samples prior to testing. For instance, ASTM B117 [48] is the industrial standard for salt fog conditioning. The standard requires sodium chloride in a slightly acidic solution (pH = 6.5 to 7.2) be sprayed on specimens held above ambient temperatures (308 K). Again, the test evaluation is based on the PV manufacturer's requirements.

Metals are integrated into the module as electrical conduits. The polymers encase these electrical components in the form of encapsulants and coatings. Therefore, the diffusion profile of salt through these polymers is of primary concern for material selection. A failure is denoted with increased resistance in the electrical conduit and visual indication of the onset of corrosion when the polymeric encapsulants and coatings are cut away from the metal's surfaces.

However, it should be noted that salts can also be corrosive to polymers [49]. Depending on the polymer chemistry, the salt can act as a plasticizer or initiate decomposition.

Ionomers are sold with counterions bound to the polymeric chains. When free salts are introduced from the surrounding environment, a cation exchange can occur. The observed behavior depends on the chemistry of the cation, but commonly there is a decrease in mechanical strength, making the ionomer more susceptible to permanent deformation.

Polymer degradation caused by salt corrosion is a common failure mode known to the electronics industry. When first introduced in the late 1950s, thermoplastics used for electrical sheaths had a warranty for 30 years but only lasted 10 years due to unforeseen degradation mechanisms catalyzed by salts. A small amount of water ingress allows an ionic salt to diffuse into the polymer. In the presence of electrical stress, this decreased the dielectric strength of the polymer and promoted dielectric breakdown. The phenomenon was called water-treeing and has been extensively investigated by a number of academic and industrial investigators [50–52]. In response, polymer manufacturers now modify thermoplastics with anti-water-treeing additives and chemical cross-links to decrease the polymer's susceptibility to degradation.

PV modules have the same warranty period as these electrical applications, but most modules have not been in the field long enough to verify that water-treeing of polymer insulators will not occur. A polymer's susceptibility to this form of degradation can be assessed by measuring changes in electrical properties after salt fog exposure or by measuring the polymer's chemical signature to verify it has undergone oxidative degradation.

2.5.2.3 Snow Loading

Module manufacturers perform ASTM E1830 [53] testing to simulate mechanical stresses endured during use. This testing standard simulates twist, cyclic, and static loads caused by typical environmental stresses, such as wind, snow, and ice. Peak loads of 1 to 5 kilopascals (kPa) are applied to the assembled modules during this testing. The evaluation for suitable performance is at the discretion of the PV manufacturer. Often a visual inspection is sufficient to determine if there has been a compromise in the packaging integrity.

The most common failure mode for modules under snow load is damage to the aluminum frame. Flat modules are typically installed at an angle to maximize light collection. Due to the angle of installation, ice and snow slide to the edge of the module creating an uneven load distribution. In addition, during heat and thaw cycles, water can collect in empty spaces in the frame and distort the shape. This is a complicated failure mode not captured in this ASTM standard and underlines the importance of performing outdoor testing parallel to laboratory evaluations.

2.6 Electrical Properties

Polymers can be classified as polar or nonpolar based on their chemical structure. A polar polymer has partial charges on the molecular chain that create a dipole moment. When halogens (7A elements, such as fluorine), ketones (R_1-CO-R_2), acids (R_1-COOH), alcohols (R_1-OH), or esters (R_2-COO-R_1) are included in the repeat unit, the chain will exhibit a dipole moment. Relevant to this discussion, polyesters, such as polyethylene terephthalate, have a dipole at the ester linkage (Figure 2.34). A nonpolar polymer does not exhibit a dipole moment. For instance, polyolefins, such as polyethylene, are composed of nonpolar methylene groups (R_1-CH$_2$-R_2).

Polymers are commonly called dielectrics, referring to the alignment of the polymer chains with the applied electric field (Figure 2.35). The polymer chains will shift to create an internal electric field that balances the charges of the applied external field. The effectiveness of a polymer to balance the applied charge is a method used to characterize the polymer's electric performance as an insulator.

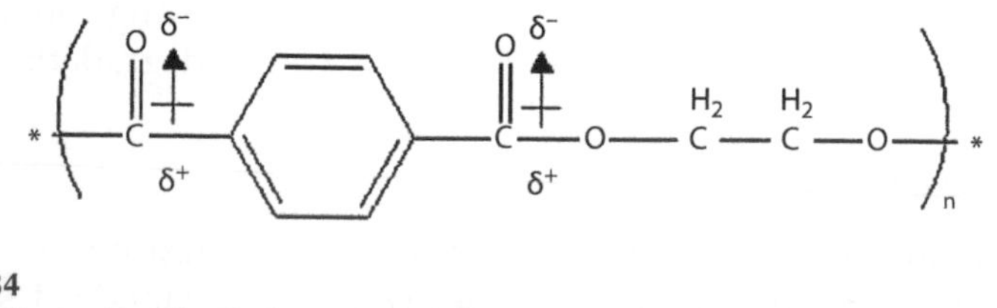

FIGURE 2.34
Chemical polarity in polyethylene terephthalate.

2.6.1 Material Properties: Dielectric Properties

Dielectric constant, dielectric strength, and volume resistivity are related material properties commonly reported by polymer manufacturers. All three measurements are an indication of the polymer's ability to create electric charge stabilization in an electric field.

The relative dielectric constant is the ratio of the polymer's capacitance to that of air. It is a measure of how well the polymer separates the charges on a capacitor. All polymers have a dielectric constant larger than one, indicating they provide better charge separation and stabilization than air. A polar polymer (e.g., polyethylene terephthalate) will have a dielectric constant between three and ten, while for nonpolar (e.g., polyethylene) polymers, the dielectric constant is typically less than three (Table 2.7) [54,55]. Polar polymers are better at charge stabilization due to the dipole moment in their chain. When placed between two capacitor plates, polar polymer chains organize to neutralize the plate's charges. If the polymer is nonpolar, the electrons around the atoms reorient to create charge neutrality.

The observed dielectric properties are highly dependent on the experimental conditions. When using alternating current (AC), it is standard to report the voltage frequency used for the measurement. Polar polymers have the highest sensitivity to the applied frequency. At higher frequencies,

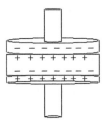

FIGURE 2.35
The charge separation in a polymer sample balances the charges on the surrounding capacitor plates.

TABLE 2.7

Polymers and Their Corresponding Dielectric Constant

Polymer Name	Frequency	Dielectric Constant
Polyethylene—low density	60 Hz	2.7
Polyvinyl fluoride	60 Hz	10
Polyethylene terephthalate	60 Hz	3.7
Ethylene vinyl acetate copolymer	1 kHz	2.8

Sources: Data from J.A. Brydson, 1999, *Plastics Materials*, 7th Ed., Oxford: Butterworth-Heinermann; J. Frados, 1976, *Plastics Engineering Handbook of the Society of the Plastics Industry*, New York: SPI.

the polymer will not have time to align with the applied charges, and the measured dielectric constant will be slightly lower.

An insulator is defined by a high dielectric strength. Dielectric strength is the breakdown voltage as a function of specimen thickness, measured in units of volts per millimeter (V/mm). Electrical breakdown occurs when the polymer begins to chemically decompose, manifesting in polymer flow or a burn-through hole. Therefore, an effective insulator stabilizes high voltages without compromise to the polymer's shape or integrity.

Dielectric strength is typically measured in accordance with ASTM D149 [56]. The setup requires a polymer sample sandwiched between two flat, metal electrodes. One of three test methods—the short time, slow-rate-of-rise, or step-by-step method—is used to apply voltage across the sample. The ASTM procedure allows for either alternating current (AC) or direct current (DC) voltage. DC voltage is the relevant type for PV applications.

The short time procedure requires a voltage increase at a specified rate, commonly 500 volts per second (V/s), until dielectric breakdown occurs [57]. The slow-rate-of-rise and step-by-step methodologies use the results obtained during the short time experiment in their procedure.

The slow-rate-of-rise method starts at 50% of the short time breakdown voltage, and the voltage is increased continually until breakdown occurs. The most common rate of rise is 100 V/s [57].

The step-by-step method starts at 50% of the breakdown voltage found in the short time method. The voltage is increased at either 20-, 60-, or 300-second intervals until dielectric breakdown occurs. A 60-second interval is the most popular method in North America [57].

The polymer thickness can influence the observed dielectric behavior. When dielectric strength is graphed as a function of thickness, the slope is defined as the intrinsic dielectric strength. A good insulator will have a shallow slope, indicating the dielectric strength is insensitive to the sample thickness. When a steep slope is observed, there is a strong dependence on sample dimensions. In general, thinner polymer samples will be more susceptible to electrical breakdown due to the shorter distance between electrodes. To avoid this variability, standard sample thicknesses, between 0.5 and 4 millimeters (mm), are used for all three test methods.

Volume resistivity is a common industrial metric to measure the polymer's insulating properties. During the test, the polymer is placed between two plates set apart at a distance of 6.35 centimeters (cm), as specified in ASTM D991 [58]. A DC voltage is applied, and resistance is calculated from the current that passes through the system. The volume resistivity equals the product of the polymer's surface area (A), in square centimeters (cm²), and the measured resistance (R), in Ohms (Ω), divided by the distance (d), in centimeters, between the electrodes (Equation 2.29):

$$\text{Volume Resistivity} = \frac{R(A)}{d} \tag{2.29}$$

TABLE 2.8

Underwriters Laboratories Comparative Tracking Index
(CTI) Assignments Based on Observed Tracking Voltage

Tracking Voltage (V)	Performance Level Categories
>600	0
599–400	1
399–250	2
249–175	3
174–100	4
<100	5

Source: Data from UL 746A "Polymeric Materials—Short-Term
Property Evaluations," November 1, 2000, Underwriters
Laboratories, Inc., Camas, WA, www.ul.com.

Values are reported in Ohm–centimeters ($\Omega \bullet$ cm), with higher values indicating better insulating properties.

The addition of ionic permeants into a polymer typically increases the polymer's electrical conductivity and thereby decreases its insulating properties. The UL comparative tracking index (CTI) tests a polymer's susceptibility to this mode of electrical breakdown. The CTI is an assigned index based on the voltage required to cause electrical tracking in the polymer when 50 drops of 0.1 weight percent (wt%) ammonium chloride solution are placed on the polymer surface (Table 2.8). A higher CTI voltage is a better insulator. Most PV applications require a CTI performance level category of two or below.

Various polymer additives included in the encapsulant formulation will alter the observed CTI voltage. For instance, higher concentrations of some ultraviolet (UV) additives (e.g., carbon black) and flame retardants can lower the CTI voltage, while mineral fillers increase it. Therefore, there is a potential trade-off between weathering, flammability, mechanical, and electrical requirements when selecting a formulation.

2.6.2 Photovoltaic Module Performance

Electrical testing for PV modules is dictated by UL 1703 [30]. The document provides guidance on electrical performance and durability testing. Wet-leak and high-potential testing are the two most common electrical tests performed by module manufacturers.

2.6.2.1 Wet-Leak Testing

Wet leak is a performance test used to verify manufacturing quality. The assembled module is placed in water containing a surfactant to increase the water's inherent electrical conductivity. The module is left to soak a minimum of 2 to 5 minutes to verify the integrity of the seal around the electrical components. The voltage is kept constant at 1000 V while the current is monitored over a span of 60 seconds. There must be no drift in current larger than 20 micro-amps (µA) for the module to pass UL 1703. A failing performance is most commonly the result of a compromise in the packaging integrity that exposes live electrical components.

2.6.2.2 High-Potential Testing

High-potential (HiPot) electrical testing is performed on an assembled module to verify electrical safety. The module is placed between two electrical terminals, and voltage is slowly raised to 1000 DC volts plus twice the module rating. The current must not drift more than 50 µA for 1 minute. No flash or arcing can be observed in order for the insulation to pass the test.

It is important that the module pass HiPot testing directly after manufacture and throughout the duration of the service life. A failure directly after processing is usually the result of manufacturing error. As an example, if a processing step is missed and electrical components are not fully encapsulated, the module may fail HiPot testing. Polymeric insulators are also susceptible to chemical breakdown in the field due to environmental exposure. Polymeric degradation can lead to impaired electrical resistance. Therefore, simulated aging tests should be performed on the prototype prior to product launch.

In service, there are multiple stresses placed on the module. Humidity, light, and temperature cycling identical to the geographic location with the highest module installation are commonly duplicated for durability testing. Modules are typically cycled under these conditions and then removed for HiPot testing. However, because diffusion profiles through a polymer can be altered when an electric potential is present, environmental conditioning should also be performed under electrical load. Therefore, many manufacturers perform constant stress tests below the breakdown voltage while the module is cycled in an environmental chamber.

In addition, it is important that the entire assembly be tested, because polymers are susceptible to mechanical creep in the layered material stack [59]. The weight of the superstrate glass on EVA encapsulants has been reported as a cause of electrical breakdown between cell leads when the polymer thinned as a result of mechanical creep during thermal cycling.

2.7 Flammability

Ignition, combustion, and flame propagation are the three principal processes used to define flammability. Ignition is the mechanism of fire creation between a fuel and oxidizer. As it is confined to this discussion, ignition is caused by an external heat source (i.e., a flame). Combustion is the chemical degradation of the material into gaseous products and char. Flame spread is a complex phenomenon related to a number of environmental variables, such as fuel concentration, wind speed, temperature, and mechanical slope. It generally describes the flame propagation along a surface.

When flammability is characterized during product certifications, a complex phenomenon of ignition, combustion, and flame propagation occurs. The observations are a combination of intrinsic and extrinsic experimental factors. For instance, the polymeric components of the product and the wind speed during testing both influence the test outcome.

Despite this ambiguity, the extensive use of polymers in consumer products has forced regulatory groups to develop a methodology for predicting flammability. Specifically, in the recent decades, lightweight polymers have been extensively used in automotive and aeronautic applications to replace heavier, metallic components. Reduced vehicle weight has increased fuel efficiency; however, the industry needs to verify passenger safety after material substitutions. As the governing body for passenger transportation in the United States, the Department of Transportation (DOT) performed considerable research on the correlation of material properties and flammability [60]. Heat release capacity has been the only material parameter successfully identified as an intrinsic flammability metric for polymers. The correlation has been so good that the U.S. Federal Aviation Administration (FAA) has regulated polymer flammability by setting heat release capacity limits for polymers used in passenger cabins.

2.7.1 Material Properties

Most polymer manufacturers report flammability ratings conducted in accordance with UL 94 [61]. The test involves placing a flame against a polymer held at a specified orientation. The duration of the burn and its propagation over the surface are the basis for the flammability rating (Table 2.9). The highest rating a polymer can receive is 5VA, indicating the flame self-extinguishes with minimal damage to the polymer. The lowest rating a polymer can have is HB.

These empirical measurements can be theoretically predicted using additive molar group theory. The theoretical estimates solely require the polymer's chemical formula; therefore, this theoretical approach does not include other additives in the commercial formulation. In Chapter 3, commercial formulations of polymers will be discussed in greater detail. Relevant to this discussion, it is important to understand that commercial formulations contain a number of small molecule additives to suppress the polymer's inherent flammability. Due to time and cost constraints, it is impossible for PV design engineers to empirically evaluate all potential materials. Though limited, these theoretical calculations are a method for thinning a material selection list to verify the right polymer classifications are considered for an application.

TABLE 2.9

UL 94 Ratings and Descriptions

Flammability Rating	Testing Specifics
5VA	The long axis of the sample is oriented vertically above the flame tip. During a surface burn, the polymer stops burning within 60 seconds after removing the flame. There is no burn-through after five applications of a flame with each application lasting 5 seconds.
5VB	The long axis of the sample is oriented vertically above the flame tip. During a surface burn, the polymer stops burning within 60 seconds after removing the flame. There is a burn-through after five applications of a flame with each application lasting 5 seconds.
V-0	The long axis of the sample is oriented vertically above the flame tip. The polymer self-extinguishes within 10 seconds after removing the flame.
V-1	The long axis of the sample is oriented vertically above the flame tip. The polymer self-extinguishes. Burning stops within 60 seconds after removing the flame.
V-2	The long axis of the sample is oriented vertically above the flame tip. Burning stops within 60 seconds after removing the flame. Flames can drip on the cotton below and start a fire.
HB	The long axis of the sample is oriented horizontally to the flame tip. Flames burn at a rate less than that specified in the standard.

TABLE 2.10

Selected Structural Group, Molar Mass, and Heat Release
Capacity Related to Ethylene Vinyl Acetate (EVA)

Structural Group	Molar Mass (M_i) (g/mol)	Molar Contribution to Heat Release Capacity (ψ_i) (kJ/(mol • K))
$-CH_2-$	14	16.7
$-CH-$	13	26.6
$-O-C=O$	44	−39.5
$-CH_3$	15	22.5

Source: Data from R. Walters, R.E. Lyon, September 2001, "Calculating Polymer Flammability from Molar Group Contributions," DOT/FAA/AR-01/31.

Since 1968, additive molar group theory has been used to predict the thermal behavior of small molecules. The original theory required a molecular property be considered as the combined effect of each atom. This theory has been modified to include the atomic connectivity of polymers by using structural groups as the smallest subdivision. This theoretical approach requires the polymer chains be further subdivided from repeat units into structural groups. For instance, the ethylene vinyl acetate (EVA) copolymer repeat units are composed of methyl, methine, methylene, and ester groups (Table 2.10, Figure 2.36). The heat release capacities (ψ_i) for various structural groups have been compiled in charts documented by the U.S. Department of Transportation (DOT).

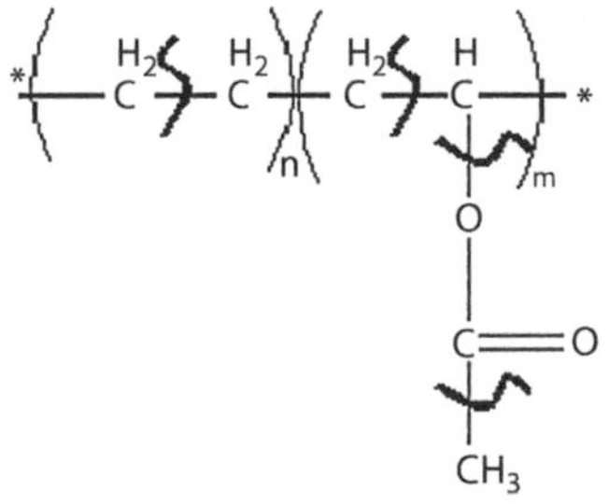

FIGURE 2.36
Chemical structure of ethylene vinyl acetate (EVA) copolymer with the structural groups isolated.

The structural group molar mass (M_i) is the summation of the atomic molar mass contributions. For instance, a methyl group (CH_3) is composed of three hydrogen atoms and one carbon atom. Using the periodic table, hydrogen has a molar mass of 1 gram per mole (g/mol) and carbon is 12 g/mol, so the molar mass of a methyl group is 15 g/mol.

The heat release capacity (η_c) on a mass basis can be determined by the sum of the product of the number of identical chemical groups (N_i) and the molar group contribution to the heat release capacity divided by the sum of the product of the number of identical chemical groups (N_i) and the molar mass (M_i) of the group (Equation 2.30).

$$\eta_c = \frac{\sum_i N_i \psi_i}{\sum_i N_i M_i} \tag{2.30}$$

Example calculation for EVA copolymer:

The numerator ($\sum N_i \psi_i$): (3) (16.7) + (1) (−39.5) + (1) (22.5) + (1) (26.6)
= 59.7 kJ/(mol • K)
The denominator ($\sum N_i M_i$): (3) (14) + (1) (44) + (1) (15) + (1) (13)
= 114 g/mol

$$\eta_c = \frac{\sum_i N_i \psi_i}{\sum_i N_i M_i} = \frac{59.7\ \text{kJ/(mol} \bullet \text{K)}}{114\ \text{g/mol}} \left(\frac{1000\ \text{J}}{1\text{kJ}} \right) = 524\ \text{J/(g} \bullet \text{K)}$$

Heat release capacity is correlated to the UL 94 flammability rating. If the heat generated by the polymer during combustion is not enough to sustain flame propagation, the polymer is designated as self-extinguishing. Empirically, this behavior occurs when the polymer's heat release capacity is below 200 J/(g • K), resulting in a V-0 rating. In contrast, when polymers have a heat capacity greater than 400 J/(g • K), they hold enough energy to sustain flame propagation without the presence of an independent heat source. Above 400 J/(g • K), polymers have an HB rating. Values between 200 and 400 J/(g • K) have an intermediate designation [60].

Again, it is important to remember this approximation only considers the polymeric structure. Returning to EVA, commercial resins sold by DuPont under the trade name Elvax®, have a V-0 rating because they contain flame retardants in the formulation to suppress the polymer's inherent flammability. These additives increase the material costs and can be minimized by choosing a polymer with inherent flame retardant characteristics.

When a polymer is used to encapsulate live electrical parts, there are a number of practical tests performed by UL to verify the formulation has the required integrity to withstand the operating conditions. The two most widely used in the PV industry are high-current arc ignition (HAI) and hot wire ignition (HWI). This is not a comprehensive analysis of all flammability testing performed by UL for electrical components. Therefore, the reader is encouraged to consider his or her polymer application and contact UL for a complete list of tests required for product certification.

Under specification UL 746A [62], the high-current arc ignition (HAI) refers to the number of electrical arcing events applied to the surface of a polymer before its ignition. A rating is designated by the mean number of arcs applied (Table 2.11). A common specification for polymeric encapsulants and pottants is a performance-level category (PLC) of 2 or 3.

A similar test is hot wire ignition (HWI) described in ASTM test standard D3874 [63]. Under these test conditions, a polymer is placed next to an ignition source. Typically, the polymer is wrapped around a live wire. If the polymer fails to ignite, the number of seconds required for burn-through is recorded. The time for ignition or burn-through defines the HWI rating (Table 2.12).

TABLE 2.11

UL 746A Category Assignments Based on
Number of Arcs Observed in the Test

Mean Number of Arcs	Assigned Rating/Performance Level Category (PLC)
>120	0
119–60	1
59–30	2
29–15	3
<15	4

Source: Data from UL 746A "Polymeric Materials— Short-Term Property Evaluations," November 1, 2000, Underwriters Laboratories, Inc., Camas, WA, www.ul.com.

TABLE 2.12

ASTM D3874 Category Assignment Based on Mean Time to Ignition Observed in the Test

Mean Time to Ignition (Seconds)	Assigned Rating/Performance Level Category (PLC)
120 and longer	0
119–60	1
59–30	2
29–15	3
14–7	4
<7	5

Source: Data from ASTM D3874-10 "Standard Test Method for Ignition of Materials by Hot Wire Sources," 2004, ASTM International, West Conshohocken, PA, 1997, DOI: 10.1520/D3874-10, www.astm.org.

2.7.2 Photovoltaic Module Performance

Erring on the side of conservatism, PV manufacturers would only integrate polymeric materials with a flame rating of V-0, or better. However, the required flame rating for product certification is dependent on the definition and function of the components. To illustrate this point, junction boxes will be used as an example. Based on IEC 61730, the junction box is commonly defined as an enclosure that houses the electronic components. The electronic components are further encased by a polymeric pottant that acts as a weathering barrier to the delicate metal circuits and wires. If the polymer is part of an enclosure, a flame rating of 5VA is typically required. If it is a pottant inside the enclosure, a flame rating of HB is commonly considered acceptable.

Ultimately, it is the flammability of the entire product, tested against UL 790 [64], that dictates certification compliance. UL 790 includes both Spread of Flame and Burning Brand tests. Both are performed while the module is exposed to a wind speed of 12 miles per hour (mph). In order to receive the classification corresponding to the test conditions, the module must not slip from its test position or produce flying brands during testing. The Spread of Flame test identifies the module's susceptibility toward flame propagation when an impinging flame, of specified temperature and duration, is applied to the surface (Table 2.13). The classification is based on the distance the flame spreads across the module. Alternatively, the Burning Brand Test specifies burning wood as a heat source. The wood is placed on the PV module to determine which components are susceptible to ignition. The size and weight of the burning brand define the testing classification (Table 2.14). The test concludes when the brand is consumed and all flames are extinguished. Based on the performance of the entire product, the product data sheet for a module will include a designation of fire classification:

TABLE 2.13

UL 790 Spread of Flame Test Classification Based on Flame
Temperature, Test Duration, and Maximum Flame Spread

Classification	Flame Temperature (K)	Test Duration (minutes)	Maximum Spread (meters)
A	1033	10	1.8
B	1033	10	2.4
C	977	4	4.0

Source: Data from UL 790 "Standard Test Methods for Fire Tests of Roof Coverings," April 22, 2004, Underwriters Laboratories, Inc., Camas, WA, www.ul.com.

TABLE 2.14

UL 790 Burning Brand Test Classifications Based on the Number,
Size, and Weight of Brands

Classification	Number of Brands per Test	Size of Brands (meters × meters)	Weight of Each Brand (grams)
A	1	0.30 × 0.30	1995.8
B	2	0.15 × 0.15	496.1
C	20	0.013 × 0.013	9.1

Source: Data from UL 790 "Standard Test Methods for Fire Tests of Roof Coverings," April 22, 2004, Underwriters Laboratories, Inc., Camas, WA, www.ul.com.

Class A, B, or C. Class A is the highest rating, and C is the lowest. In all cases, the module and its components will ignite and combust; however, the higher rating gives occupants a better opportunity to exit before the building is consumed.

Historically, flat modules with EVA encapsulants receive the lowest class rating. Combustible vapors (e.g., methane and ethane gas) released from decomposing EVA result in rapid flame propagation. These highly flammable volatile organics atomize under extreme heat and fall onto adjacent roofing materials resulting in flame spread. PV manufacturers have solved this problem with two approaches. First, they decreased the concentration of flammable small molecules in the module. EVA formulations have been reengineered to eliminate highly volatile organics. In addition, additives have been included in the formulation to increase the thermal stability of EVA. Second, they increased the burst strength (34 to 344 kPa from 422 to 588 K) of the backsheet in order to contain the flammable molecules inside the module [59]. Fluorinated polymeric backsheets were integrated into flat-module packaging because they have a high flame rating [59]. Both of these requirements have added cost to the packaging materials.

2.8 Weathering Stability

Weathering describes the physical and chemical processes of material degradation caused by environmental exposure. These processes create changes in a number of the properties discussed in the previous sections [65–67]. Most importantly, photovoltaic manufacturers must accurately predict the drift in mechanical and optical properties in order to guarantee a performance warranty.

Due to the long warranty of PV modules (25 to 30 years), the time to market for new products does not allow for outdoor weathering for the same duration as the product lifetime. PV manufacturers must gain confidence in their service life predictions by performing accelerated tests.

Predicting weatherability is dependent on accurately simulating the conditions found in an outdoor environment. Humidity, rain, temperature, and ultraviolet irradiance must be duplicated with the same frequency, intensity, and duration as the geographical climate of interest [68]. In order to validate indoor testing, a typical approach is to perform accelerated testing in parallel with outdoor exposure to verify the findings in the simulated environment [69].

Prior to the late 1970s, the majority of PV modules were used in aerospace applications. These highly engineered applications commanded the use of high-value polydimethylsiloxane encapsulants known for their resistance to harsh environmental conditions. Superior thermal, oxidative, and UV resistance is imparted to the polymer by the strength of the silicon–oxygen (Si-O) bonds in the chain's structure and the cross-linked polymeric morphology. However, as PV modules moved toward terrestrial commercialization, EVA copolymers became an industrial favorite due to their lower cost. Due to this change in polymers, the life expectancy data of polydimethylsiloxane encapsulants could not be utilized for these new terrestrial applications. Without this historical data, the industry was forced to perform accelerated testing to predict product lifetimes. Despite this "due diligence" testing, a number of manufacturers surveyed in 1993 admitted they did not realize EVA was susceptible to discoloration until their modules were returned. This confusion resulted from an imperfect knowledge of available literature and execution of unrealistically accelerated experiments [70].

To the author's knowledge, there are no books written specifically on the weathering of PV packaging. However, by 1993, numerous technical articles and government publications reported yellowed EVA after exposure to thermal stress and UV light. Therefore, a literature search should have revealed this as a probable failure mechanism even if the manufacturers did not directly observe it [70].

Even though EVA discoloration was reported, most authors did not expect this to be an in-service failure mode. Unrealistic thermal acceleration temperatures and inappropriate approximations caused them to overpredict

EVA's life expectancy. Experiments included a thermal acceleration factor generated at an upper temperature limit of 150°C (423 K). This temperature was chosen to maximize the acceleration factor and decrease the experimental time. However, this temperature is unrealistic from a polymer perspective, because it is above the EVA melt temperature (318 to 379 K). In addition, the general rule of thumb, "chemical reaction rates double for every increase in 10°C," was used to estimate the acceleration factor in reliability reports. This generalization rarely fits empirical polymeric data. Neither condition is acceptable for predicting polymeric degradation. This section includes mechanisms and governing equations accepted in the polymer industry to predict weathering behavior.

2.8.1 Material Properties

Weatherability is an inherent property correlated to the polymer structure and formulation. Weathering causes polymer bonds to break, manifesting in a macroscopic change in material properties. Energy must supercede a threshold to break the chemical bonds in a polymer chain. Once the chemical bonds are broken, the changes in molecular weight will alter the observed physical properties. Therefore, commercial polymeric formulations include small molecular additives to shield the polymer chains from harmful energy. Both the polymeric structure and the formulation additives must be considered during the experimental design and data analysis.

PV packaging materials must be transmissive to light while simultaneously acting as a barrier to oxygen, moisture, and other environmental permeants. These barrier properties are a critical performance parameter, because PV cells can undergo hydrolytic and oxidative degradation. Therefore, a principal concern is to choose polymers with a low moisture and oxygen transmission rate. However, it is equally important to realize that polymers are also susceptible to hydrolytic and oxidative degradation [71]. Therefore, both the polymer's transmission rates and its susceptibility to moisture- and oxygen-induced degradation should be included in material evaluation procedures.

2.8.1.1 Stabilizer Package

Polymers are susceptible to photolytic degradation in the presence of UV light. Ultraviolet describes the wavelengths of light between x-rays and visible light in the electromagnetic spectrum. The UV spectrum is divided into three components based on wavelength: UV-C (100 to 280 nm), UV-B (280 to 320 nm), and UV-A (320 to 400 nm). While other components exist, our discussion will be limited to these, because they have the most significant influence on a polymer's stability.

Photolytic degradation is a chemical reaction catalyzed by the absorption of high-energy light. The chemical bonds in the polymers break apart creating a free radical. A free radical is a highly reactive chemical species with at

TABLE 2.15

Chemical Bonds, Polymers, and Sensitive Wavelengths

Chemical Bond	Polymers with These Bonds	Sensitive Wavelength (nm)
C=O	Polyester, ethylene vinyl acetate (EVA)	164
C-H (methane)	Epoxy, EVA	280
Si-O	Polydimethylsiloxane	321
C-C (aliphatic)	Epoxy, Polyester, EVA	345

Sources: Data from C. Trust, 2001, *Recent Advances in Environmentally Compatible Polymers: Cellucon '99 Proceedings,* Cambridge: Woodhead; A. Davis, D. Sims, 1983, *Weathering of Polymers,* London: Applied Science.

least one unpaired electron. The free radicals can react with other polymer chains forming a cross-linked network or within the same chemical chain causing a decrease in the polymer's molecular weight [72].

The chemical bonds constituting polymer chains have varying susceptibilities to UV radiation. For example, the aliphatic, carbon–carbon (C-C), bonds found in polyesters, epoxies, and EVA are susceptible to higher wavelengths of light than the silicon–oxygen bonds in polydimethylsiloxanes (Table 2.15).

To protect the polymer chains from harmful radiation, commercial thermoplastic formulations contain small molecular additives called UV stabilizers. UV stabilizers are composed of chemicals categorized by their mode of interaction with UV radiation. The four general classifications include screeners, absorbers, quenching agents, and free radical scavengers.

UV screeners remove radiant UV light from the surrounding polymer chains by reflecting or absorbing incident radiation (Figure 2.37). Metals are excellent reflectors, but they are often not compounded into polymer formulations because they increase the polymer's electrical conductivity. Magnesium oxide, calcium carbonate, and barytes are excellent reflectors of 300 nm to 400 nm light. Zinc oxide and titanium dioxide are the most common UV screeners, because they are less abrasive to processing equipment than the aforementioned compounds. However, these compounds are relatively poor reflectors. The most effective and therefore widely used screener is carbon black. The efficiency of carbon black is highly dependent on particle size and loading. Small particle sizes on the order of 10 nm to 20 nm are the most effective at loadings between 2 and 5 wt%. The exact chemical mechanism to describe how carbon black creates UV stability is still debated. However, it is hypothesized that carbon black both absorbs harmful UV radiation and terminates free radicals formed during degradation [75].

UV absorbers absorb harmful UV radiation and release it in the form of heat before it can initiate degradation of the polymer chains. There are four chemical classes: oxanilide, hydroxybenzophenone, hydroxyphenyl-s-triazine,

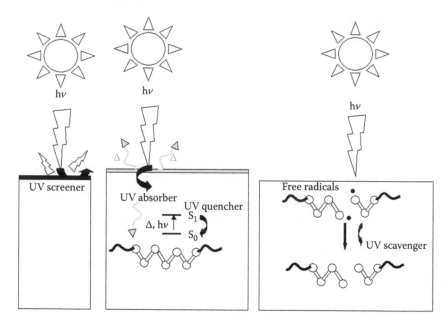

FIGURE 2.37
Depiction of the ultraviolet (UV) stability methods of (I) screeners, (II) absorbers, quenchers, and (III) scavengers.

TABLE 2.16

Chemical Class of Ultraviolet Absorber and UV Absorption Range

UV Absorber	Start of Absorption (nm)
Oxanilide	320
Hydroxybenzophenone	330
Hydroxyphenyl-s-triazine	350
Benzotriazole	360

and benzotriazole. Each classification has a slightly different UV absorption. The UV light not absorbed by the additive is transmitted through to the underlying PV cell. Oxanilide has the largest transmission of UV light, absorbing wavelengths lower than 320 nm, and benzotriazole has the smallest UV light transmission, absorbing wavelengths lower than 360 nm (Table 2.16) [73,74].

Quenching agents, commonly metal chelates, also absorb harmful radiation before it can cause bond rupture in polymer chains [72]. Once absorbed, quenchers dissipate the energy through an electronic transition. Their electronic transition typically results in visible light emission giving the polymer a slight coloration.

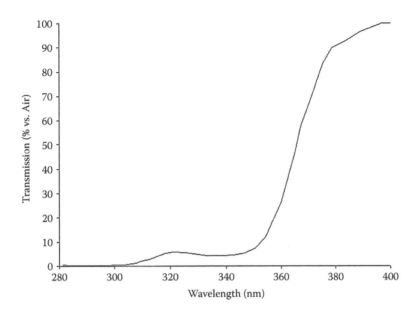

FIGURE 2.38
Typical ultraviolet-visible (UV-Vis) spectrum for hindered amines light stabilizers (HALS).

Hindered amine light stabilizers (HALS) are a special class of organic compounds called free radical scavengers. They are formulated to terminate free radicals that occur during photolytic degradation [76]. All HALS are organic compounds that contain a secondary amine group constrained within a hydrocarbon ring. There are a number of proposed chemical mechanisms to describe HALS stabilization, but most proposed theories involve a HALS reaction with the polymeric free radicals generated during photodegradation [77]. This prevents the free radical from creating further polymeric breakdown. The presence of HALS is denoted in the polymer's UV-Vis spectrum as a double absorption. Peak maxima usually occur at 300 and 350 nm, but the exact wavelength is dependent on the commercial grade of HALS in the polymer (Figure 2.38). Those absorbed wavelengths will not reach the underlying PV cells and therefore cannot be harnessed for electricity generation.

2.8.1.2 Transmission Rates

Both transmission rate and permeability describe the physical phenomena of a permeant moving through a medium. Transmission rates define the concentration of permeant through a unit area after a specified time at a constant temperature and humidity. Permeability defines the concentration of permeant through a unit area and sample thickness after a specified time at a constant temperature and humidity. Transmission rates are commonly

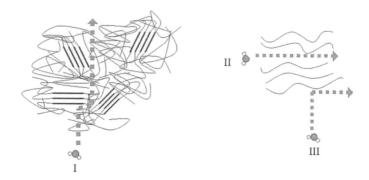

FIGURE 2.39
Permeant path through (I) semicrystalline polymers, (II) parallel to chain orientation, and (III) perpendicular to chain orientation.

used in industry, while permeability is commonly used in academic literature [78]. The two concepts are interchangeable as long as the experimentalist defines the thickness of the sample when transmission rates are reported.

Environmental conditions alter the recorded transmission rates. Increased temperature and humidity typically increase the mobility of most permeants through the polymer. In both cases, on a microscopic level, the polymer chains separate, making it easier for a permeant to move through the morphology.

Percent crystallinity and degree of orientation are similar morphologies that have very different effects on transmission rates. Crystals are typically randomly placed throughout the thickness of the polymer sample. The transmission rate decreases because the crystals' random placement increases the tortuosity of the permeant's path through the polymer [79] (Figure 2.39). Similarly, polymer chains can align when molded into a final shape during processing. Transmission rates increase when measured in the same direction of the chain alignment. The permeant can easily navigate between the oriented chain morphology. However, when measured perpendicular to the orientation, the permeant cannot easily move through the chains, thereby decreasing the transmission rate. A number of researchers have developed characteristic equations to quantify and predict these relationships [80]. Their descriptive analysis is beyond the scope of this section. However, the important point is that processing conditions can influence the observed transmission rates. Experiments on the final configuration are more instructive than the polymer samples received from the supplier or the reported properties cited on the data sheet.

2.8.1.2.1 Water Vapor Transmission Rate

Water vapor transmission rate (WVTR) is synonymous with moisture vapor transmission rate (MVTR). WVTR is a material property that describes

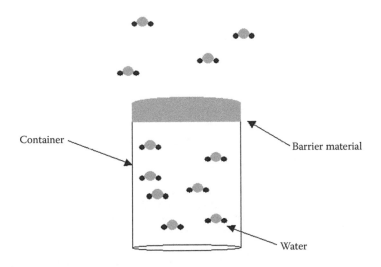

FIGURE 2.40
Depiction of a water vapor transmission rate (WVTR) measurement.

the transmission rate of moisture through a polymeric substrate at a controlled external temperature and humidity. This property is a vital packaging consideration when a hermetic seal is required to ensure product performance.

WVTR can easily be evaluated by placing a reservoir of water in a beaker and sealing it with the test substrate (Figure 2.40). The entire system is placed in a humidity chamber so both temperature and humidity can be simultaneously controlled and monitored. The change in the water weight (Δw), measured in grams (g), in the reservoir as a function of time (t), measured in days, and divided by the barrier film's surface area (A), measured in square meters (m^2), gives the water vapor transmission rate (Equation 2.31. A WVTR is typically reported in grams transmitted per meter squared per day (g/m^2/day).

$$WVTR = \frac{\Delta w}{tA}$$

(2.31)

WVTR is typically performed at 85°C/85% relative humidity (RH) (358 K/85% RH). There are a number of standards, such as ASTM F1249 [81] and ASTM E96 [82], that outline these measurements. Commercial instruments that adhere to these standards are also available. As an example, Mocon® (Minneapolis, Minnesota) markets a number of WVTR instruments with different sensitivity thresholds sold under the model names Permatran® and Aquatran®.

Ionomers and polydimethylsiloxanes have some of the smallest WVTR values (Table 2.17) [83]. Based on these polymer characteristics, they are logical

TABLE 2.17

Polymer Class and Water Vapor Transmission
Rates (WVTRs)

Polymer Name	WVTR (g/m²/day)
Polyethylene-*b*-polymethacrylic acid salt-*b*-polymethylacrylate	0.3
Polyethylene terephthalate	3.4
Ethylene vinyl acetate copolymer	33–27
Polydimethylsiloxane	0.01–0.006

Source: Data from G.D. Barber, G.J. Jorgensen, K. Terwilliger, S.H. Glick, J. Pern, T.J. McMahon, 2002, New Barrier Coating Materials for PV Module Backsheets, *Proceedings of the 29th IEEE PV Specialists Conference.*

choices for packaging. However, the WVTRs reported in this table and those reported by polymer manufacturers are based on test films. Because a WVTR measurement is dependent on the exposed surface area to the surrounding environment, it is imperative for solar manufacturers to perform an evaluation on their modules during material selection [84,85].

WVTR describes the polymer's barrier properties. However, the polymer can also absorb and retain moisture from the surrounding environment. This absorbed water can act as a plasticizer, significantly changing the polymer's mechanical, electrical, and optical properties. For this reason, water absorption capacity is a common material consideration in polymer science. Polymer handbooks will report the equilibrium water capacity for various commercial polymers.

Water absorption capacity experiments are simpler than WVTR measurements. Briefly, the polymer's absorption capacity is measured by comparing the polymer's weight in a dry state and after prolonged exposure to a controlled humidity. The sensitivity for these measurements necessitates the use of microbalances to identify small changes in the sample's weight. Commercial moisture analyzers possess a microbalance in a confined, controlled environment. The sample is placed onto a preweighed pan and heated over a user-specified temperature ramp. The moisture driven off the sample is calculated by the change in weight after reaching the final drying temperature. The percent change in weight is the equilibrium moisture capacity. These instruments can be purchased from various analytical vendors, such as Mettler-Toledo (Greifensee, Switzerland), Ohaus (Pine Brook, New Jersey), and Denver Instruments (Bohemia, New York).

If a polymer with a high water absorption capacity is used for packaging applications, it must be dried before encapsulation. If it is not dried, the encapsulant will release water into the adjacent PV cells during diurnal thermal cycling. This is a concern for polar polymers, like ionomers, which have a high absorption capacity. Nonpolar polymers, like polydimethylsiloxanes, will be relatively insensitive to changes to the surrounding humidity and will not require drying. To avoid drying, most manufacturers target a water absorption capacity of 0.2 to 0.3 wt%. Based on the data presented in Table 2.18 [86,87], this requirement limits the class of polymeric encapsulants to some polydimethylsiloxanes and EVA copolymers.

2.8.1.2.2 Oxygen Transmission Rate

Oxygen transmission rates (OTRs) are evaluated with similar principles as described for water vapor transmission. However, as the name implies, the permeant is oxygen rather than water. Most commercial instruments adhere to ASTM D3985 [88]. They have a reservoir of oxygen gas separated by the test membrane from a reservoir of nitrogen gas (Figure 2.41).

TABLE 2.18

Encapsulation Polymers and Equilibrium Water Absorption Capacity

Polymer Name	Water Absorption Capacity (wt%)
Polyethylene-*b*-polymethacrylic acid salt-*b*-polymethylacrylate	29–11
Polyethylene terephthalate	0.5
Ethylene vinyl acetate copolymer	0.13–0.005
Polydimethylsiloxane	0.4–0.1

Sources: Data from J.E. Mark, 1999, *Polymer Data Handbook*, Oxford: Oxford University Press; P.F. Bruins, 1970, *Silicone Technology*, New York: Interscience.

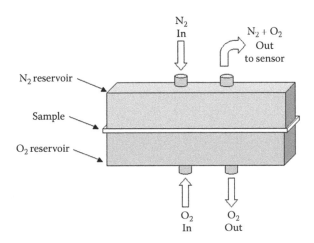

FIGURE 2.41
General depiction of the major components of an oxygen transmission rate (OTR) apparatus.

The nitrogen is constantly analyzed with a coulometric sensor to detect the concentration of oxygen that diffuses through the membrane into the adjacent reservoir. The OTR is measured in volume of diffused oxygen per surface area of membrane per unit of time, typically cubic centimeters per meters squared per day ($cm^3/m^2/day$). The temperature and humidity are controlled and reported with the measurements. Commercial instruments are sold by Mocon Industries under the trade name OX-TRAN® and Illinois Instruments (Johnsburg, Illinois) marketed as 8000 series analyzers.

2.8.1.3 Hydrolytic Degradation

Hydrolytic degradation is a chemical mechanism that describes material deterioration in the presence of water. The susceptibility of polymers to this form of degradation is dependent on their chemical structure. Some polar polymers have a chemical structure prone to hydrolytic degradation. For instance, polyesters have an ester (R_1-COO-R_2) chemical linkage that creates a polarity in the chemical chain. During hydrolysis, the polymer's molecular weight decreases as the ester linkage along the polymer backbone reacts with ingressing water molecules (Figure 2.42).

Degradation is typically monitored as a change in mechanical properties rather than chemical changes in chain structure. Properties are measured before and after long-term humidity exposure, 1000 hours of 85% RH and 85°C (358 K). The service lifetime is estimated as the exposure time required for the mechanical properties to decrease below the failure requirement, described in Section 2.5.1.

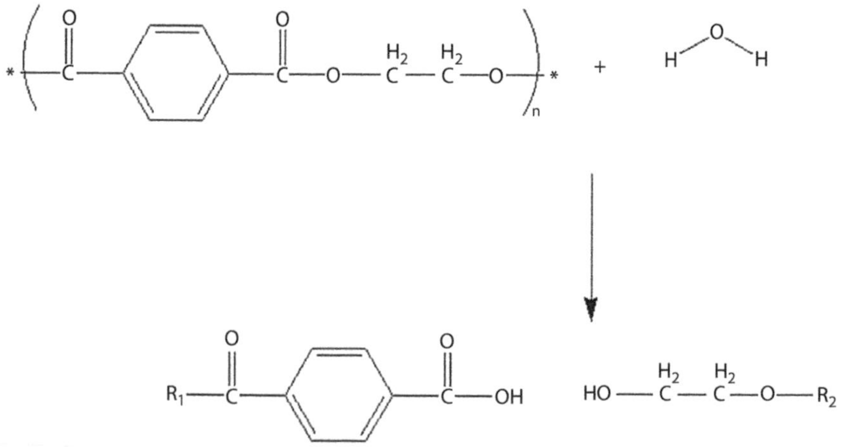

FIGURE 2.42
Hydrolytic degradation of a polyethylene terephthalate chain.

Martin and Gardner reported humidity degradation experiments on 32 different thermoplastic and thermoset polymers with varying degrees of humidity exposure, from 10 months to 3 years [89]. They found the lifetime for hydrolysis-sensitive polymers doubles for every quarter reduction in environmental humidity. In addition, two mechanical properties, elongation to break and impact resistance, were the most sensitive parameters for predicting lifetimes. There are a number of similar studies available in the academic literature to help design engineers predict a polymer's susceptibility to hydrolysis [90,91].

2.8.1.4 Oxidative Degradation

Oxidative degradation describes the chemical mechanism by which polymer chains react with oxygen, changing their length and altering their macroscopic material properties. This chemistry is catalyzed in the presence of heat, light, or certain aggressive chemicals, such as acids, bases, and metal oxides.

The polymer's chemistry and formulation dictate its susceptibility to oxidative degradation. Polymers with a carbon backbone in their chemical structure are the most susceptible to oxidative degradation. Under extreme heat, light, or chemical agents, the carbon–hydrogen (C-H) bonds can break, forming free radicals; this is described as initiation (Figure 2.43) [92]. During propagation reactions, the free radical can react with molecular oxygen forming highly reactive peroxide radicals, or it can propagate between different small molecules and molecular chains. Ultimately, the reaction will terminate causing a decrease in chain length, branching, or cross-links.

To increase the polymer stability against free radicals, small molecule additives, called antioxidants, are added to commercial formulations. The additives react with the oxygen, effectively blocking their reaction with the polymer chains. Fluorinated polymers and polydimethylsiloxanes have stronger molecular bonds making them less susceptible to this form of degradation and requiring no antioxidants in their formulation.

The susceptibility of a polymer formulation to oxidative degradation is measured by the oxidative induction time using differential scanning calorimetry (OIT-DSC). This technique uses the same instrumentation as previously described for DSC. The sample temperature is ramped to an equilibrium temperature above the highest polymer thermal transition, and the atmosphere is changed to oxygen. When the polymer begins to degrade, heat will be released. This is designated by an exotherm on the spectrum. The dwell time between reaching the equilibrium temperature and the onset of the exotherm is recorded as the oxidative induction time (OIT). The more resistant the polymer formulation to oxidative degradation, the longer is the OIT.

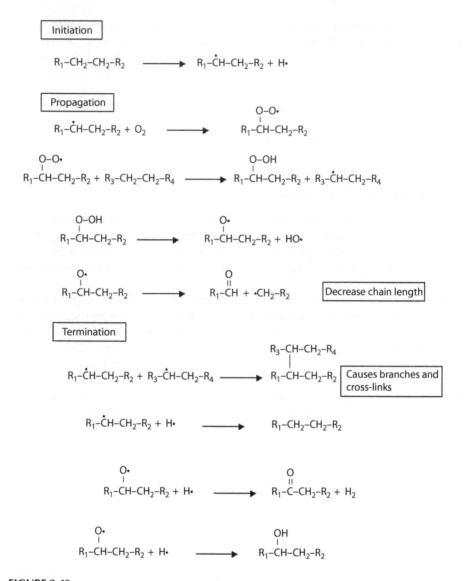

FIGURE 2.43
Oxidative degradation of a polyethylene chain.

2.8.2 Photovoltaic Module Performance

Designing a product to be deployed in all global climates and have a 25- to 30-year lifetime is an unprecedented requirement for polymer packaging [93,94]. There is limited practical knowledge on the service life of commercial-grade polymers in all geographical climates. Furthermore, there is no universal equation to predict the degradation kinetics of all polymers in all environments. Instead, each commercial formulation must be tested on a case-by-case basis.

As noted earlier, there has been significant confusion in the PV industry concerning how to accurately predict the service life of polymeric encapsulants. The following sections survey experimental techniques and predictive modeling used in polymer science to make material selections for outdoor applications.

2.8.2.1 Accelerated UV Aging Techniques

For a PV manufacturer, there are two concerns relevant for this discussion. First, manufacturers are concerned whether the UV additive package will inhibit light collection and harm module performance. Second, they worry UV exposure to the polymers will affect its properties and potentially reduce the service life of the module.

Polymer compounders and manufacturers purchase UV additives to be included in the commercial polymeric formulations. The specific UV additives used in a polymer's formulation are unknown to consumers and considered a trade secret among polymer manufacturers. Although this detail in the chemical formulation will not be provided in the technical data sheet, the consumer can identify the presence of some additives simply by looking at the polymer. As an example, in some cases, carbon black will make the polymer black, and talc will make the polymer opaque. These UV screeners cannot be used in polymeric components that will shade the underlying PV cells. Therefore, they are acceptable UV stabilizers for the frame components but not for the encapsulants.

Optically clear polymers may not necessarily perform well as an encapsulant. The UV additives should be optimized to minimize interference with the PV cell's collection of light. Encapsulants should maximize transparency to those wavelengths of highest internal quantum efficiency (IQE) for the PV cell. Some polymer manufacturers have product literature that recommends various polymeric grades for different types of PV cell chemistries. Regardless, it is advisable to obtain a UV-Vis spectrum of potential encapsulant candidates prior to prototype development. The UV-Vis spectrum overlaid with the IQE curve insures that there is no usable light screened out by the UV additives in the encapsulant. Using a silicon cell as an example, the UV additives in an EVA formulation cut off a small amount of transmission below 350 nm (Figure 2.44). In comparison, the polydimethylsiloxane encapsulant does not contain UV stabilizers that cut off usable UV light from the PV cell.

If the inherent optical properties are considered acceptable, then the candidate materials move to durability assessments. At a minimum, polymer candidates are tested in accordance with UL 746C [95]. This standard assigns one of two categories (i.e., f1 and f2) to a polymer based on its weathering

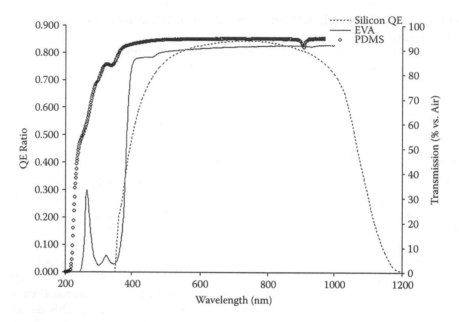

FIGURE 2.44

Ultraviolet-visible (UV-Vis) transmission spectra of polydimethylsiloxane (Dow Corning RTV 615) and ethylene vinyl acetate (EVA) (DuPont Encapsolar® 135) encapsulants overlaid with a single crystalline silicon cell quantum efficiency curve.

performance. The test requires sample conditioning, under UV, moisture, and heat, prior to property evaluation. The UV exposure under a carbon source must be performed for 720 hours or under xenon-arc source for 1000 hours. In both cases, there must also be water exposure for 7 days at 70°C (343 K). Mechanical impact, strength, and flammability must not significantly change after exposure to receive a passing qualification of f1. When a polymer is denoted as f2, it either has not met these requirements or has not been fully tested.

A 20-year service life is a rigorous outdoor requirement not guaranteed with an f1 rating. From a formulation perspective, polymers that have an f1 rating do not necessarily have the proper UV package to last 25 to 30 years in all global climates. Therefore, most PV manufacturers choose to perform their own weathering predictions to verify their products meet their specified warranty.

The first consideration in performing a service life prediction is to determine the chemical additive or material property responsible for optimal product performance (Figure 2.45).

If it is known a priori that the presence of additives is required for the retention of the critical material property, then the change in concentration of these additives (e.g., UV additive) can be monitored as a function of

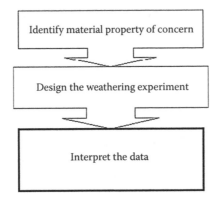

FIGURE 2.45
Decision flowchart for developing an accelerated weathering test.

accelerated testing. For instance, the end of the polymer's service life could be denoted by the depletion of the UV additive as identified by an increase in the UV transmission. The time to failure (t_{fail}), in seconds, is derived from the rate constant for UV additive depletion (k), in inverse seconds, the transmission of the sample prior to aging (T_o), and the timed dose of UV light or accelerated stimulus required for material failure (D_{fail}), in seconds. The dose required for failure is derived from a test specimen with no additives (Equation 2.32) [96]:

$$t_{fail} = \frac{1}{k}\log\left[\frac{10^{kDfail} + T_o - 1}{T_o}\right]$$

(2.32)

This is a common technique used for thin polymer samples, such as coatings where the loss of UV additives will result in a drastic change in material properties.

Rarely will a design engineer have the necessary chemical knowledge to perform an analysis based on changes in chemical formulations. It is more common for a critical material parameter to be measured as a function of the accelerated stimulus. For instance, changes in the yellowness index affect the critical function of encapsulants. When this parameter is simultaneously monitored as a function of outdoor exposure and accelerated UV testing, the data can be extrapolated to the failure point in order to predict a service life.

2.8.2.2 Developing an Accelerated Test

Once the material parameter has been selected, the accelerated test must be designed. As mentioned earlier, design engineers require a method to realistically accelerate these conditions in a laboratory and extrapolate those results to expectations in an outdoor environment. Weathering is a complicated chemical phenomenon caused by a combination of stresses, such as light, temperature, moisture, and soiling. Even so, a number of researchers proposed equations to describe the relationship between indoor and outdoor weathering.

In 1966, Kamal presented the Exposure Parameter Approach that predicts an exponential correlation between a material parameter (y) and the exposure time (t) (Equation 2.33). The constants A, B, and C are determined by fitting the data from accelerated testing. Once the coefficients are determined, the failure limit of the material parameter can be inserted into the equation (y), and the service life is predicted by solving for time (t) [97]:

$$y = A \exp\left[B(t - C)\right] \qquad (2.33)$$

Howard and Gilroy proposed another correlation in 1969. It directly correlated artificial exposure time (t_a) to outdoor weathering (t_n) through the fitting constants B and k [98,99] (Equation 2.34):

$$t_a = B + t_n^k \qquad (2.34)$$

Finally, Fischer and Ketola proposed that artificial exposure time (t_2) is equivalent to outdoor weathering (t_1) multiplied by an acceleration factor (AF) [100] (Equation 2.35):

$$t_1 = AF t_2 \qquad (2.35)$$

In this expression, the acceleration factor is a function of light, heat, water, and soiling. Experimental conditions and equations to identify the functional form of the acceleration factor are the focus of the remainder of this discussion.

Weathering protocols can provide a guideline for test development. ASTM standards are used in the polymer industry to measure UV stability (Table 2.19) [101–106]. These standards are excellent templates for product development. For instance, they can be used to screen various polymeric formulations for PV applications. However, these standards must be modified to have the specificity required for making service life predictions.

Artificial light sources do not accurately simulate natural light over the entire electromagnetic spectrum. Comparison of the artificial light to that of the AM 1.5 allows the design engineer to determine if the acceleration factor generated by these results is going to overestimate or underestimate the real-world exposure. There are five principal light sources commercially available for solar simulators: xenon arc, UV-A fluorescent lamp, UV-B

TABLE 2.19

ASTM Protocol and Corresponding Experimental Details

ASTM Protocol	UV Source and Experimental Details
ASTM D1435-05 Standard Practice for Outdoor Weathering of Plastics	• Outdoor testing without concentration • Standard is best for a relative comparison between plastic performance in a single geographical location • Suggests at least a 12-month exposure and no extrapolation of results for service life predictions
ASTM D2565-99 Standard Practice for Xenon-Arc Exposure of Plastics Intended for Outdoor Applications	• Xenon-arc source • Includes the effect of light, moisture, and heat • Specifically designed for plastics
ASTM D4329-05 Standard Practice for Fluorescent UV Exposure of Plastics	• Fluorescent light source • Includes the effect of light, moisture, and heat • Specifically designed for plastics
ASTM D4364-05 Standard Practice for Performing Outdoor Accelerated Weathering Tests of Plastics Using Concentrated Sunlight	• Fresnel lens reflectors for concentrated light • Attempts to simulate stress conditions in the desert and subtropical climates
ASTM G90-05 Standard Practice for Performing Accelerated Outdoor Weathering of Nonmetallic Materials Using Concentrated Natural Sunlight	• Fresnel lens reflectors for concentrated light • Focused on the conditions and parameters of the apparatus which must be controlled for reproducible data
ASTM G155-05a Standard Practice for Operating Xenon-Arc Light Apparatus for Exposure of Non-Metallic Materials	• Xenon-arc sources • Simulates rain or humidity exposure during ultraviolet (UV) stress

fluorescent lamp, metal halide, and carbon arc. Xenon-arc light sources most closely match the UV area of the electromagnetic spectrum. Therefore, this source more closely predicts polymeric properties after natural light exposure.

Once the light source is chosen, the exposure duration must be considered. Typically, either the harshest expected environment or the environment with the highest number of installations is simulated. The annual UV exposure dose for various geographies can be found on government agencies' and manufacturer's Web sites. Within the United States, these include the National Renewable Energy Laboratory (NREL), the National Oceanic and Atmospheric Administration (NOAA), and Atlas Material Testing Solutions (Table 2.20) [107]. The warranty period multiplied by the average annual UV exposure gives the total UV dose required for the test. The total dose divided by the lamp intensity gives the exposure duration for the simulated test.

TABLE 2.20

Annual Exposure of Ultraviolet (UV) and Ultraviolet-Visible
(UV-Vs) Light in Select Geographies

Region	Coordinates	400 to 295 nm UV (MJ/m^2)	800 to 295 nm UV + Visible (MJ/m^2)
Southern Florida 26°S	25°52'N 80°27'W	390	4400
Central Arizona 34°S	33°54'N 112°8'W	440	5200
Southern Europe 45°S	43°8'N 5°49'E	324	4300
Western Europe 45°S	51°57'N 4°10'E	270	2850

Source: Data from O. Haillant, 2010, Personal Communication, Atlas Material
Testing Technology GmbH, Linsengericht, Germany.

Note: 1 MJ/m^2 = 0.2778 kWh/m^2 = 277.8 Wh/m^2, where kW/m^2 is kilo-Watt-
hours per square meter, and Wh/m^2 is Watt-hour per square meter.

Example Duration Calculation:

The annual UV dose in central Arizona: 440 MJ year^{-1} m^{-2}
For a 25-year equivalent exposure: 25 years * 440 MJ year^{-1} m^{-2} = 11,000
MJ m^{-2}
The experimental lamp produces: 0.00005 MW m^{-2}
Using the conversion 1 W = 1 Js^{-1}
This is 0.00005 MJ m^{-2} per second or 0.18 MJ m^{-2} per hour
11,000 MJ m^{-2}/0.18 MJ m^{-2} hr^{-1} = 61,111 hours ≈ 7.0 years
This is in terms of light hours, meaning the lamp is continuously on. If the lamp
is cycled, the total hours of testing need to be adjusted accordingly.

The total expected exposure should consider the angle of the module's
installation. The angle of exposure of the plastic component in the module
should be duplicated during testing. These angles of exposure are relevant,
because two types of light rays are incident on the polymer surface: direct
and diffuse rays. Direct rays are incident at a 90-degree angle with the surface. Diffuse rays are reflected rays from other surfaces or those rays that
form an angle less than 90 degrees from the polymer's surface. When the
specimens are parallel to the ground and directly exposed to overhead sunlight, the highest amount of direct rays are incident, resulting in the largest amount of irradiance. When specimens are perpendicular to the ground,
mostly diffuse rays are incident on the surface, and therefore the total irradiance decreases. Taking Southern Florida exposure (295 to 385 nm) as an
example, a module perpendicular with the ground will receive 180 MJ/m^2,
while a module parallel to the ground will receive 310 MJ/m^2 [108].

Light and dark cycles are included in most ASTM standards because
they allow time for UV additives to regenerate. This chemistry is most

effective at the surface, but it can be removed through evaporation and rainfall. This depletes the chemistry, causing additional molecules to diffuse up to the surface from the bulk. Depending on the thickness of the polymeric part, it is possible to deplete the entire reservoir of additives prior to reaching the end of warranty for the product. In this case, the polymer chains will be left susceptible to UV-induced degradation. For this reason, it is important to include appropriate temperatures, humidity, and rainfall cycles to insure the design is robust enough to function for the expected service life.

The effect of soiling on the polymer's UV stability is a concern for material selection for outdoor applications. Gubanski and Wankowicz investigated the UV absorption of various soiling agents, such as methylcellulose, graphite, and carbon black [109]. They found increasing UV absorption with various soiling agents. This behavior suggests soil and soot collection on a polymer's surface act as a UV screener. For a transparent polydimethylsiloxane elastomer sprinkled with Arizona dust, there is a decrease in light transmission over all wavelengths, with the largest decreases occurring over the visible and UV spectral regions (Figure 2.46). Therefore, the absence of these environmental conditions during testing will lead to incorrect extrapolations of acceleration factors. Artificial soiling cycles are not included in the cited ASTM weathering standards. Although it is accepted that soiling will

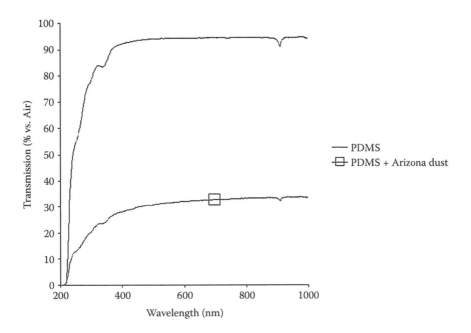

FIGURE 2.46
Change in ultraviolet-visible (UV-Vis) transmission of a polydimethylsiloxane (Dow Corning RTV 615) after it is sprinkled with Arizona dust (average 50 micrometer [µm] diameter).

impact UV degradation, there have been no proposed methods to universally predict the magnitude of this effect.

It is imperative to perform testing on the same polymers formed from the same processing techniques that will be used in the final design. Not all EVA grades are chemically and functionally equivalent. For instance, each grade of EVA has a proprietary UV package. Therefore, different grades will not weather at the same rate. To add further complication, the polymer's morphology can change during processing, affecting the weathering rate [110]. As an example, processing conditions can create residual stress in the part. This increased orientation allows free radicals to freely propagate, increasing the rate of UV-induced chain degradation.

The importance of measuring the exact polymer composition and configuration is important and merits reiteration. Pern and Glick illustrated this point with their research results presented at the National Center for Photovoltaics (NCPV) Program Review Meeting [111]. They assembled crystalline silicon PV cells with various encapsulants and placed them in accelerated exposure tests. At high concentrations, 9x Suns and 145°C (418 K), there was no drop in EVA transmission but a drop in cell efficiency. The same effect was seen with polydimethylsiloxane encapsulants. The authors attributed their findings to the migration of deleterious chemical additives from the polymer package to the PV cell interface, but the migrant chemicals were never identified. This study powerfully illustrates two important concepts. First, the PV industry is comfortable with higher temperatures and UV concentrations than deemed acceptable in the polymer industry. For instance, in the polymer industry, 2 to 4x Suns and temperatures below the polymer's transitions are common. Second, the assembled product should be evaluated during prototype development in order to observe other potential failure mechanisms only identified in the completed assembly.

2.8.2.3 Data Analysis

Once the data have been collected, they must be analyzed to identify the acceleration factor (AF). The graph of the material property (e.g., yellowness index, YI) as a function of time (t) will fit a kinetic expression described with a degradation rate constant (k_{deg}). Kinetic expressions are discussed in detail in chemical textbooks. They are defined by the dependence of the rate expression on the concentration of chemical reactants included in the degradation mechanism. The reader is referred to these primary texts for more explanation concerning rate constants.

To facilitate this discussion, a simple zero-order rate constant will be used as an example. The change in the yellowness index is equal to the product of the degradation rate constant (k_{deg}) and time (t) (Equation 2.36):

$$\Delta YI = -k_{deg}t \qquad (2.36)$$

The rate constants observed over a series of experimental conditions (e.g., temperatures) will typically fit one of three principal models: the Arrhenius, the Erying, or the Inverse Power Law. The Arrhenius equation solely includes the effect of temperature on the degradation kinetics. A plot of the natural log of the degradation rate constant (k_{deg}) and inverse temperature (T), in Kelvin, will give a straight line with the characteristic activation energy (E_a), in electronvolts (eV), divided by the Boltzman constant ($k_b = 8.617385 \cdot 10^{-5}$ eVK^{-1}), as the slope. The interecept is the natural logarithm of the material constant (A) (Equation 2.37):

$$\ln(k_{deg}) = \ln A - \frac{E_a}{k_b T} \qquad (2.37)$$

The Eryling equation is a combination of temperature (T) and one additional environmental stress (S_1). The degradation rate constant (k_{deg}) is graphed as a function of stress, while the activation energy, A, and B_1 and C_1 are all constants used to describe the relationship (Equation 2.38):

$$k_{deg} = AT^\alpha \exp\left[-\frac{E_a}{k_b T} + \left(B_1 + \frac{C_1}{T} \right) S_1 \right] \qquad (2.38)$$

The Inverse Power Law describes nonthermal accelerated stress. The degradation rate constant is equivalent to the stress level (V) raised to the model parameter (γ) multiplied by another model parameter (A) (Equation 2.39):

$$k_{deg} = AV^\gamma \qquad (2.39)$$

A logrithmic plot of the degration rate constant versus the logrithim of stress gives a slope of γ and an intercept of the logrithm of A (Equation 2.40):

$$\log k_{deg} = \log A + \gamma \log V \qquad (2.40)$$

A value of zero for the model parameter (γ) indicates stress does not influence the degradation.

Any one of these rate equations that best describes the observed material behavior can be used as the acceleration factor (AF) (Equation 2.35). During weathering experiments, a number of stresses combine to affect the degradation. Therefore, it is more common to insert the Erying or the Inverse Power Law as the acceleration factor. However, for simplicity, the Arrhenius equation is used as an example. The change in the performance variable (ΔP) is the product of the degradation rate constant and the exposure time in the artificial weatheirng experiment (t_2) (Equation 2.41):

$$\Delta P(t_2) = \Delta YI = -k_{\text{deg}}t_2 = -At_2 \exp\left(-\frac{E_a}{k_b T_2}\right) \tag{2.41}$$

The same equation can be written for outdoor weathering substituting outdoor weathering exposure duration (t_1) for accelerated exposure (t_2) and the mean outdoor temperature (T_1) for the accelerated temperature (T_2). Assuming the same change in performance is observed between both tests, the two equations can be set equal to derive the Arrhenius form (Equation 2.42) of the acceleration factor presented in Equation (2.35) [112]:

$$t_1 = t_2 \exp\left\{-\frac{E_a}{k_b}\left(\frac{1}{T_2} - \frac{1}{T_1}\right)\right\} = AFt_2 \tag{2.42}$$

Using this equation, the service life (t_1) can be predicted based on the time to failure in the accelerated test (t_2), the experimental derived activation energy, the temperature of the accelerated test, and the outdoor service temperature.

Alternative theories have been offered that discount the importance of multiple, simultaneous stresses during light exposure. This assumption drastically simplifies service life predictions.

Reciprocity theory was first introduced by Bunsen and Roscoein in 1859 based on their experimental observations of the chemicals used for photography development. Reciprocity has been widely accepted in photobiology, photoconduction, and photography.

Reciprocity theory states a photoresponse is solely dependent on the absolute quantity of irradiance (I); therefore, the product of the irradiance and exposure time (t) is constant (Equation 2.43) [113]:

$$I \bullet t = \text{constant} \tag{2.43}$$

This provides a method for exposing the polymer at high intensities and constant irradiance and then extrapolating to the anticipated response at actual use conditions. It implies the outdoor irradiance (I_1) mutilpied by the exposure time (t_1) is equivalent to the indoor simulated irradiance (I_2) mulitplied by exposure time (t_2) (Equation 2.44):

$$I_1 t_1 = I_2 t_2 \tag{2.44}$$

Schwarzschild's law is a more popular law of reciprocity. It was proposed in 1900 to allow for a better fit to the data at extreme levels of radiant flux. The coefficient p is dependent on the material and the conditions of the tests (Equation 2.45). However, when $p = 1$, the equation collapses to the general law of reciprocity.

$$I^p \bullet t = \text{constant} \tag{2.45}$$

It is important to note that these equations were first proposed and applied to small molecules and not polymer chains. The polymer's additional chemical complexity is a primary reason these simplistic equations have yet to receive general acceptance in polymer science. However, they are offered here to provide a well-rounded representation of weathering theory.

References

1. Cuddihy, E.F.; Baum, B.; Willis, P. 1979. Low-Cost Encapsulation Materials for Terrestrial Solar Cell Modules. *Solar Energy* 4: 389–396.
2. Arya, R. 2004. Technology and Market Challenges to Mainstream Thin-Film Photovoltaic Modules and Applications. Arya International, Inc., Williamsburg, VA.
3. European Parliament. 2003. Directive 2002/95/EC of The European Parliament and of the Council of 27 January 2003 on the Restriction of the Use of Certain Hazardous Substances in Electrical and Electronic Equipment. *Official Journal of the European Union* 46: 19–23.
4. Kanter, J. November 8, 2009. Balancing Energy Needs and Material Hazards. http://www.nytimes.com/
5. European Parliament. 2003. Directive 2002/96/EC of the European Parliament and of the Council of 27 January 2003 on Waste Electrical and Electronic Equipment (WEEE). *Official Journal of the European Union* 46:24–38.
6. Underwriters Laboratory. 2010. Using UL Recognized Components in Your PV Modules. http://ul.com/photovoltaics.
7. Fink, D.G.; Beaty, H.W. 1987. *Standard Handbook for Electrical Engineers*, 12th Ed. New York: McGraw-Hill.
8. International Electrotechnical Commission. 2010. About the IEC. http://www.iec.ch/
9. IEC 61730-1. "Photovoltaic (PV) Module Safety Qualification—Part 1: Requirements for Construction." October 1, 2004. International Electrotechnical Commission, Geneva, Switzerland. www.iec.ch
10. IEC 61730-2. "Photovoltaic (PV) Module Safety Qualification—Part 2: Requirements for Testing." October 1, 2004. International Electrotechnical Commission, Geneva, Switzerland. www.iec.ch
11. American Standard for Testing Materials. 2010. About ASTM International. http://www.astm.org/

12. Brandrup, J.; Immergut, E.H.; Grulke, E.A. 1999. *Polymer Handbook*. New York: Wiley.
13. Mark, J.E. 1996. *Physical Properties of Polymer Handbook*. Woodbury, NY: AIP Press.
14. NuSil Silicones. 2007. LS-3354 Optically Clear Encapsulation Gel Product Profile. http://www.silicone-polymers.ie/
15. ASTM Standard D1925-70. 1988, Withdrawn 1995. "Standard Test Method for Yellowness Index of Plastics." ASTM International, West Conshohocken, PA. www.astm.org
16. Karwowski, W. 2001. *International Encyclopedia of Ergonomics and Human Factors*, Vol. 1. London: Taylor & Francis.
17. Harris, D.C. 1997. *Exploring Chemical Analysis*. New York: W.H. Freeman.
18. Cuddihy, E.F.; Coulbert, C.D.; Liang, R.H.; Gupta, A.; Willis, P.; Baum, B. 1983. Applications of Ethylene Vinyl Acetate as an Encapsulation Material for Terrestrial Photovoltaic Modules DOE/JPL-1012-87.
19. ASTM Standard E1036. 2002. "Standard Test Methods for Electrical Performance of Nonconcentrator Terrestrial Photovoltaic Modules and Arrays Using Reference Cells." ASTM International, West Conshohocken, PA. DOI: 10.1520/E1036-08, www.astm.org
20. Tsuno, Y.; Hishikawa, Y.; Kurokawa, K. 2005. Temperature and Irradiance Dependence of the I-V Curves of Various Kinds of Solar Cells. 15th International Photovoltaic Science and Engineering Conference (PVSEC-15). Shanghai, China, 26-1:422–423.
21. Markvart, T.; Castaner, L. 2003. *Practical Handbook of Photovoltaics: Fundamentals and Applications*. Oxford: Elsevier Science.
22. Pern, F.J.; Glick, S.H. 2000. "Effects of Accelerated Exposure Testing (AET) Conditions on Performance Degradation of Solar Cells and Encapsulants." Program and Proceedings of the NCPV Program Review Meeting.
23. Hernández-Sánchez, F. 2007. Heat Capacity Measurement in Polymers Using a Differential Scanning Calorimeter: Area Measurement Method. *Journal of Applied Polymer Science* 105: 3562–3567.
24. Ballara, A.; Trotignon, J.P.; Verdu, J. 1986. Skin-Core Structure of Polyetheretherketone Injection-Moulded Parts from DSC Measurements. *Journal of Materials Science Letters* 5: 706–708.
25. ASTM Standard D696-08. 1998. "Standard Test Method for Coefficient of Linear Thermal Expansion of Plastics Between –30°C and 30°C with a Vitreous Silica Dilatometer" 2003. ASTM International, West Conshohocken, PA. DOI: 10.1520/D0696-08, www.astm.org
26. ASTM Standard D5470-06. 1996. "Standard Test Method for Thermal Transmission Properties of Thermally Conductive Electrical Insulation Materials" 2001. ASTM International, West Conshohocken, PA. DOI: 10.1520/D5470-06, www.astm.org
27. UL 746B Revision 5. "Standard for Safety of Polymeric Materials-Long Term Property Evaluations." Underwriters Laboratories, Camas, WA. www.ul.com
28. IEC 61730-1. October 1, 2004. "Photovoltaic (PV) Module Safety Qualification—Part 1: Requirements for Construction." International Electrotechnical Commission, Geneva.

29. IEC 61730-2. October 1, 2004. "Photovoltaic (PV) Module Safety Qualification—Part 2: Requirements for Testing." International Electrotechnical Commission, Geneva. www/iec.ch
30. UL 1703. "Flat-Plate Photovoltaic Modules and Panels." Underwriters Laboratories, Camas, WA. www.ul.com
31. Underwriters Laboratory (UL). 2010. "Using UL Recognized Components in Your PV Modules." www.ul.com
32. ASTM Standard D2240-05, 2004. 2000. "Standard Test Method for Rubber Property-Durometer Hardness." ASTM International, West Conshohocken, PA. DOI: 10.1520/D2240-05, www.astm.org
33. ASTM Standard D903-98, 2004. 1998. "Standard Test Method for Peel or Stripping Strength of Adhesive Bonds." ASTM International, West Conshohocken, PA. DOI: 10.1520/D0903-98R04, www.astm.org
34. ASTM Standard D638-10, 2008. 1999. "Standard Test Methods for Tensile Properties of Plastics." ASTM International, West Conshohocken, PA. DOI: 10.1520/D0638-10, www.astm.org
35. ASTM Standard D695-08, 2002. 1996. "Standard Test Method for Compressive Properties of Rigid Plastics." ASTM International, West Conshohocken, PA. DOI: 10.1520/D0695-08, www.astm.org
36. ASTM Standard D2990-09, 2001. 1995. "Standard Test Method for Tensile, Compressive, and Flexural Creep and Creep-Rupture of Plastics." ASTM International, West Conshohocken, PA. DOI: 10.1520/D2990-09, www.astm.org
37. ASTM Standard D395-03. 2008. "Standard Test Methods for Rubber Property-Compression Set." ASTM International, West Conshohocken, PA. DOI: 10.1520/D0395-03R08, www.astm.org
38. ASTM Standard D256-06ae1, 2006. 2006. "Standard Test Methods for Determining the Izod Pendulum Impact Resistance of Plastics." ASTM International, West Conshohocken, PA. DOI: 10.1520/D0256-06A, www.astm.org
39. ASTM D790-07e1, 2007. 2000. "Standard Test Methods for Flexural Properties of Unreinforced and Reinforced Plastics and Electrical Insulating Materials." DOI: 10.1520/D0790-07, www.astm.org
40. Whitfield, K. 2009. Accelerated Thermal Degradation of Polymers in Photovoltaic Applications. ATCAE Solar 2009 Atlas Technical Conference on Aging and Evaluation, December 8–9, Phoenix, AZ.
41. King, D.L.; Boyson, W.E.; Kratochivil, J.A. 2004. "Photovoltaic Array Performance Model," Sandia National Laboratories, document SAND2004-3535.
42. Kurtz, S.; Miller, D.; Kempe, M.; Bosco, N.; Whitefield, K.; Wohlgemuth, J.; Dhere, N.; Zgonena, T. 2009. "Evaluation of High-Temperature Exposure of Photovoltaic Modules." 34th IEEE Photovoltaic Specialists Conference, Philadelphia, PA.
43. Wohlegemuth, J.H. 2009. "Lifetime Predictions—What We Still Have to Do to Be Able to Estimate Field Performance Based on Accelerated Testing." ATCAE Solar 2009 Atlas Technical Conference on Ageing and Evaluation.
44. ASTM Standard E1171-09, 2004. 1999. "Standard Test Method for Photovoltaic Modules in Cyclic Temperature and Humidity Environments." ASTM International, West Conshohocken, PA. DOI: 10.1520/E1171-09, www.astm.org

45. SunPower. 2008. Technical Data Sheet for 225 Solar Module. www.sunpower-corp.com
46. BP Solar. 2007. Technical Data Sheet for SX 3200. http://www.bpsolar.com
47. First Solar. 2009. Technical Data Sheet for First Solar FS Series 2 PV Module. www.firstsolar.com
48. ASTM Standard B117-09, 2007. 1997. "Standard Practice for Operating Salt Spray (Fog) Apparatus." ASTM International, West Conshohocken, PA. DOI: 10.1520/B0117-09, www.astm.org
49. Doehne, E. 2002. Salt Weathering: A Selective Review. *Geological Society, London, Special Publications*. 205: 51–64.
50. Boggs, S.; Xu, J. 2001. Water Treeing-Filled versus Unfilled Cable Insulation. *IEEE* 17: 23–29.
51. Ciuprina, F.; Teissédre, G.; Filippini, J.C. 2001. Polyethylene Crosslinking and Water Treeing. *Polymer* 42: 7841–7846.
52. Dissado, L.A.; Fothergill, J.C. 1992. *Electrical Degradation and Breakdown in Polymers*. London: Peter Peregrinus.
53. ASTM Standard E1830-09, 2004. 2001. "Standard Test Methods for Determining Mechanical Integrity of Photovoltaic Modules." ASTM International, West Conshohocken, PA. DOI: 10.1520/E1830-09, www.astm.org
54. Brydson, J.A. 1999. *Plastics Materials*, 7th Ed. Oxford: Butterworth-Heinermann.
55. Frados, J. 1976. *Plastics Engineering Handbook of the Society of the Plastics Industry*. New York: SPI.
56. ASTM D149-09, 1997. 2000. "Standard Test Method for Dielectric Breakdown Voltage and Dielectric Strength of Solid Electrical Insulating Materials at Commercial Power Frequencies." ASTM International, West Conshohocken, PA. DOI: 10.1520/D0149-09, www.astm.org
57. Bartnikas, R. 1987. "Engineering Dielectrics, Vol. IIb: Electrical Properties of Solid Insulating Materials: Measurement Techniques." Baltimore: ASTM Special Technical Publication.
58. ASTM D991 (2010), 2005. 2000. "Standard Test Method for Rubber Property-Volume Resistivity of Electrically Conductive and Antistatic Products." ASTM International, West Conshohocken, PA. DOI: 10.1520/D0991-89-R10, www.astm.org
59. Willis, P.B. 1985. Investigation of Test Methods, Material Properties and Processes for Solar Cell Encapsulants. Ninth Annual Report, DOE/JPL-954527-27/85.
60. Walters, R.; Lyon, R.E. September 2001. "Calculating Polymer Flammability from Molar Group Contributions," DOT/FAA/AR-01/31.
61. UL 94. "The Standard for Safety of Flammability of Plastic Materials for Parts in Devices and Appliances Testing," October 29, 1996. Underwriters Laboratories, Inc., Camas, WA. www.ul.com
62. UL 746A. "Polymeric Materials-Short Term Property Evaluations," November 1, 2000. Underwriters Laboratories, Inc., Camas, WA. www.ul.com
63. ASTM D3874-10. 1997. "Standard Test Method for Ignition of Materials by Hot Wire Sources" 2004. ASTM International, West Conshohocken, PA. DOI: 10.1520/D3874-10, www.astm.org
64. UL 790. April 22, 2004. "Standard Test Methods for Fire Tests of Roof Coverings." Underwriters Laboratories, Inc., Camas, WA. www.ul.com

65. Hamid, S.H. 2000. *Handbook of Polymer Degradation*, 2nd Ed., Revised and Expanded. New York: Marcel Dekker.
66. Allara, D.L. 1975. Aging of Polymers. *Environmental Health Perspectives* 11: 29–33.
67. Brown, R.P. 1993. Durability of Polymer-Material Property Data. *Polymer Testing* 12: 423–428.
68. Brown, R.P. 1993. Test Procedures for Artificial Weathering. *Polymer Testing* 12: 459–466.
69. Crump, L.S. 1998. "Evaluating the Durability of Gel Coats Using Outdoor and Accelerated Weathering Techniques: A Correlation Study." Society for the Advancement of Material and Process Engineering, International SAMPE Europe Conference No19, Puteaux, France. 808: 487–499.
70. Czanderna, A.W.; Pern, F.J. 1996. Encapsulation of PV Modules Using Ethylene Vinyl Acetate a Copolymer as a Pottant: A Critical Review. *Solar Energy Materials and Solar Cells* 43: 101–181.
71. Audouin, L.; Langlois, V.; Verdu, J.; de Bruijn, J.C.M. 1994. Role of Oxygen Diffusion in Polymer Ageing: Kinetic and Mechanical Aspects. *Journal of Materials Science* 29: 1573–4803.
72. Guillet, J.E. 1972. Fundamental Processes in the UV Degradation and Stabilization of Polymers. *Pure and Applied Chemistry* 30: 135–144.
73. Trust, C. 2001. *Recent Advances in Environmentally Compatible Polymers: Cellucon '99 Proceedings*. Cambridge: Woodhead.
74. Davis, A.; Sims, D. 1983. *Weathering of Polymers*. London: Applied Science.
75. Klemchuk, P.P. 1983. Influence of Pigments on the Light Stability of Polymers: A Critical Review. *Polymer Photochemistry* 3: 1–27.
76. Klemchuk, P.P.; Gande, M.E.; Cordola, E. 1990. Hindered Amine Mechanisms: Part III—Investigations Using Isotopic Labeling. *Polymer Degradation and Stability* 27: 65–74.
77. White, J.R.; Turnbull, A. 1994. Weathering of Polymers: Mechanisms of Degradation and Stabilization, Testing Strategies and Modeling. *Journal of Materials Science* 29: 584–613.
78. Comyn, J. 1985. *Polymer Permeability*. London: Chapman & Hall.
79. Hedenqvist, M.; Gedde, U.W. 1996. Diffusion of Small-Molecule Penetrants in Semicrystalline Polymers. *Progress in Polymer Science* 21: 299–333.
80. Kempe, M. 2005. Module Encapsulant Diagnostic and Modeling. *DOE Solar Energy Technologies Program Review Meeting*, Denver, CO. NREL/CP-520-37027.
81. ASTM Standard F1249-06, 2001. 1990. "Standard Test Method for Water Vapor Transmission Rate Through Plastic Film and Sheeting Using a Modulated Infrared Sensor." ASTM International, West Conshohocken, PA. DOI: 10.1520/F1249-06, www.astm.org
82. ASTM Standard E96-00e1. 2000. "Standard Test Methods for Water Vapor Transmission of Materials." ASTM International, West Conshohocken, PA. DOI: 10.1520/E0096_E0096M-05, www.astm.org
83. Barber, G.D.; Jorgensen, G.J.; Terwilliger, K.; Glick, S.H.; Pern, J.; McMahon, T.J. 2002. New Barrier Coating Materials for PV Module Backsheets. Proceedings of the 29th IEEE PV Specialists Conference, May 14–19, New Orleans, LA.
84. Kempe, M.D. 2006. Modeling of Rates of Moisture Ingress into Photovoltaic Modules. *Solar Energy Materials & Solar Cells* 90: 2720–2738.

85. Jorgensen, G.; Terwilliger, K.; Glick, S.; Pern, J.; McMahon, T. May 2003. Materials Testing for PV Module Encapsulation NREL/CP-520-33578.
86. Mark, J.E. 1999. *Polymer Data Handbook*. Oxford: Oxford University Press.
87. Bruins, P.F. 1970. *Silicone Technology*. Easton: Interscience.
88. ASTM Standard D3985-05, 2002. 1995. "Standard Test Method for Oxygen Gas Transmission Rate Through Plastic Film and Sheeting Using a Coulometric Sensor." ASTM International, West Conshohocken, PA. DOI: 10.1520/D3985-05, www.astm.org
89. Martin, J.R.; Gardner, R.J. 1981. Effect of Long Term Humid Aging on Plastics. *Polymer Engineering and Science* 21: 557–565.
90. Apicella, A.; Nicolais, L. 1981. Environmental Aging of Epoxy Resins: Synergistic Effect of Sorbed Moisture, Temperature, and Applied Stress. *Industrial and Engineering Chemistry Product Research and Development* 20: 138–144.
91. Muzzy, J.D.; Kays, A.O. 1984. Thermoplastic vs. Thermosetting Structural Composites. *Polymer Composites* 5: 169–172.
92. Pospisil, J.; Klemchuk, P.P. 2000. *Oxidation Inhibition in Organic Materials*, Vol. 1. Boca Raton, FL: CRC Press.
93. Ito, M.; Nagai, K. 2008. Degradation Issues of Polymer Materials Used in Railway Field. *Polymer Degradation and Stability* 93: 1723–1735.
94. Berdahl, P.; Akbari, H.; Levinson, R.; Miller, W.A. 2008. Weathering of Roofing Materials—An Overview. *Construction and Building Materials* 22: 423–433.
95. UL 746C. February 12, 2001. "Polymeric Materials—Use in Electrical Equipment Evaluations." Underwriters Laboratories, Camas, WA. www.ul.com
96. Martin, J.W.; Bauer, D.R. 2002. *Service Life Prediction: Methodology and Metrologies*. Washington, DC: Oxford University Press.
97. Kamal, M.R. 1970. Cause and Effect in the Weathering of Plastics. *Polymer Engineering and Science* 10: 108–121.
98. Howard, J.B.; Gilroy, H.M. 1969. Natural and Artificial Weathering of Polyethylene Plastics. *Polymer Engineering and Science* 9: 286–294.
99. Davis, A.; Sims, D. 1983. *Weathering of Polymers*. New York: Elsevier Science.
100. Fischer, R.; Ketola, W.D. in Grossman, D.; Ketola, W.D. 1994. *Accelerated and Outdoor Durability Testing of Organic Materials*. Material Park: ASTM International. 88–111.
101. ASTM Standard D1435-05. 1999. "Standard Practice for Outdoor Weathering of Plastics." ASTM International, West Conshohocken, PA. DOI: 10.1520/D1435-05, www.astm.org
102. ASTM Standard D2565-99(2008). 1999. "Standard Practice for Xenon-Arc Exposure of Plastics Intended for Outdoor Applications." ASTM International, West Conshohocken, PA. DOI: 10.1520/D2565-99R08, www.astm.org
103. ASTM Standard D4329-05. 1999. "Standard Practice for Fluorescent UV Exposure of Plastics." ASTM International, West Conshohocken, PA. DOI: 10.1520/D4329-05, www.astm.org
104. ASTM Standard G90-05. 1998. "Standard Practice for Performing Accelerated Outdoor Weathering of Nonmetallic Materials Using Concentrated Natural Sunlight." ASTM International, West Conshohocken, PA. DOI: 10.1520/G0090-05, www.astm.org

105. ASTM Standard G155-05a, 2005. 2000. "Standard Practice for Operating Xenon Arc Light Apparatus for Exposure of Non-Metallic Materials." ASTM International, West Conshohocken, PA. DOI: 10.1520/G0155-05A, www.astm.org
106. ASTM Standard D4364-05, 2002. 1994. "Standard Practice for Performing Outdoor Accelerated Weathering Tests of Plastics Using Concentrated Sunlight." ASTM International, West Conshohocken, PA. DOI: 10.1520/D4364-05, www.astm.org
107. Haillant, O. 2010. Personal Communication. Atlas Material Testing Technology GmbH. Linsengericht, Germany.
108. ATLAS 2001. Weathering Testing Guidebook. Atlas Electric Devices Company 2062/098/200/AA/03/01.
109. Gubanski, S.M.; Wankowicz, J.G. 1989. Distribution of Natural Pollution Surface Layers on Silicone Rubber Insulators and Their UV Absorption. *IEEE Transactions on Electrical Insulation* 24: 689–697.
110. Pagan, P. 1987. The Weathering of Plastics—Photochemical Mechanisms and the Influence of Processing—Part II. *Polymers Paint Colour Journal* 177: 704–710.
111. Pern, F.J.; Glick, S.H. 2000. Effects of Accelerated Exposure Testing (AET) Conditions on Performance Degradation of Solar Cells and Encapsulants. *Program and Proceedings NCPV Program Review Meeting.* April 16–19. Denver, CO.
112. McMahon, T.J. 2004. Accelerated Testing and Failure of Thin-Film PV Modules. *Progress in Photovoltaics: Research and Applications* 12: 235–248.
113. Schwarzchild, K. 1900. On the Law of Reciprocity for Bromide of Silver Gelatin. *Astrophysical Journal* 11: 89–91.

3

Polymer Specifications for Photovoltaic (PV) Packaging and Balance of System (BOS) Components

3.1 Commercial Formulations

Commercial formulations are a combination of polymer chains and small molecular additives tailored to the consumer's requirements. Additives are included in the formulation to either enhance the polymer's inherent properties, termed modifiers, or maintain those properties under environmental stress, termed *stabilizers*. For instance, modifiers may be added to improve the polymer's inherent impact resistance, while stabilizers prevent polymers from losing impact resistance during service (Table 3.1). Both stabilizers and modifiers are requirements for polymeric, photovoltaic (PV) components.

3.1.1 Polymeric Stabilizers

Polymer degradation is a behavior-modifying chemical mechanism that can be inhibited by stabilizers. Degradation can be catalyzed by light, heat, oxygen, metal, and biological molecules. The polymer can be introduced to these catalysts during processing or service. Both organic and inorganic molecules can be used as stabilizers to deactivate these catalysts or halt degradation. Flame retardants, ultraviolet (UV) stabilizers, antioxidants, metal deactivators, and antimicrobial agents are the most relevant stabilizers for PV packaging.

Organic flame retardants are highly reactive chemical species that delay or inhibit combustion. Organic flame retardants typically contain halogen atoms. Halogens are elements from the 7A group of the periodic table and include fluorine, chlorine, bromine, and iodine. Brominated flame retardants are the most common because they have the required efficacy at the desired cost. Even though the European Union's (EU) environmental restrictions have limited the use of polybrominated biphenyls (PBB) and polybrominated diphenyl ether (PBDE), other brominated organic compounds are still widely

TABLE 3.1

Common Polymeric Additives Included in Commercial Formulations: Their Definitions and Common Classifications

Additives	Definition	Additive Classification	Chemical Classification
Cross-Linking Agents	A chemical reagent or reagents used to connect the polymer chains into a three-dimensional network	Modifier	Organic
Foaming Agents	Reactive organic molecules that increase air/void space in the polymer	Modifier	Organic or inorganic
Fillers	Inorganic particulates used to change mechanical and electrical properties as well as decrease the amount of polymer in the part	Modifier	Inorganic
Adhesion promoters	Tie molecules to improve chemical adhesion between the polymer and an adjacent substrate (e.g., glass, wood, etc.)	Modifier	Organic
Plasticizers	Molecules added to decrease the polymer's glass transition temperature	Modifier	Organic
Metal deactivator	Prevents metal catalyzed degradation	Stabilizer	Organic and inorganic
Antimicrobial agents	Molecules or polymers that inhibit microbial growth	Stabilizer	Organic
Flame retardants	Small molecular additives meant to decrease flammability and delay combustion	Stabilizer	Inorganic or organic
Ultraviolet (UV) additives	Small molecules that inhibit photolytic polymer degradation	Stabilizer	Inorganic or organic
Colorants	Pigments or colorants added to change the natural color of the polymer	Modifier	Inorganic or organic
Processing agents	Small molecular additives meant to improve stability and yield	Stabilizer or modifiers	Organic

used in polymer formulations. The halogens on the flame retardants form free radicals under heat. The free radicals decrease the polymer's flammability by accelerating the polymer's degradation and creating an impermeable char on the surface. They can also react with surrounding oxygen, decreasing its concentration and subsequently delaying combustion.

Inorganic flame retardants can also be employed to inhibit combustion. They work in one of three ways: absorbing heat away from the polymer, depleting adjacent oxygen through the release of water, or forming a protective char across the polymer's surface. Some common inorganic additives include silica, aluminum trihydrate, graphite, antimony trioxide, and magnesium hydroxide.

UV additive packages can include organic and inorganic compounds designed to either absorb or reflect harmful UV radiation or halt photolytic degradation. The different classifications of UV additives were previously discussed in Section 2.8 on weathering.

Heat stabilizers, also known as antioxidants, protect the polymer from thermal decomposition during processing. Thermal energy causes polymer chains to break and form free radicals. Once a free radical is formed, it reacts with other polymer chains to create cross-links, and it reacts with itself to reduce the polymer's molecular weight. Antioxidants react with the free radicals, effectively removing them from the chemical reaction and preventing the polymer's decomposition. They are commonly composed of organic compounds known as thiols, ascorbic acid, or polyphenols.

Metal impurities can be intentionally or unintentionally embedded in polymer formulations. Intentional sources may be formulation additives, like catalysts used in the polymerization or inorganics used as flame retardants. In addition, some applications require metal components to be encapsulated in polymers forming an interface between the two materials. Unintentional sources include ingress from environmental sources or debris from processing equipment.

Regardless of the mode of introduction, metal catalyzed degradation requires a series of coreactants including metal complexes. Oxides of zinc, lead, copper, and iron have been identified as the most problematic contaminates [1]. The metallic ion catalyzes the oxidative degradation of the polymer. Although the degradation mechanism is not fully understood, it is believed that a combination of metal, acid, water, oxygen, and heat is required. However, the reaction can be avoided by elimination of the metal catalysts. Metal chelating agents are added to polymer formulations to form metal ion complexes with the catalysts. This effectively prevents them from participating in degradation reactions.

Outdoor applications necessitate additives to protect against biological attack. Biological growth can inhibit light transmission through the superstrate glass as well as cause degradation of polymeric components. Eliminating bacterial growth requires molecules that block the bacteria's metabolic processes or rupture the microbe's cytoplasmic membrane. This requires metallic elements (e.g., silver), metallic oxides (e.g., zinc oxide or titanium oxide), or ionic chemical species (e.g., acids and ionomers) [2]. Based on the identity of these stabilizers, there can be trade-off between metal and biological stabilization. Therefore, the formulation must be carefully tailored to the expected service environment.

3.1.2 Polymeric Modifiers

Modifiers are added to alter material behavior either during processing or service. Modifiers can be used solely to improve the aesthetics but are principally added to improve mechanical behavior.

Colorants are typically only added to provide aesthetic appeal. The most popular colorants are optical brighteners used to reverse the natural yellow coloration of polyolefins (e.g., polyethylene, polypropylene). These molecules absorb UV light and emit blue light. The emission of blue light offsets the yellow and creates a white appearance. This colorant also acts as a stabilizer by removing some of the harmful UV radiation.

Cross-linking agents are added to the formulation to react with polymer chains and create a three-dimensional structure. They must not react until the final processing step when the final shape has been formed. The presence of chemical cross-links will impart either elastomeric or thermoset properties to the polymer depending on the cross-link density and polymer chemistry. These properties were extensively discussed in Chapter 1.

Plasticizers are commonly included in thermoplastics to lower the glass transition temperature, thereby increasing mechanical toughness. They are small molecules that can be depleted through natural evaporation or surface washing. Once removed, the inherent, polymeric properties are restored. Clearly, the plasticizer must remain in the polymer during the warranty period if the application requires the modified properties. Most PV applications do not use plasticized thermoplastics, because the plasticizers are susceptible to environmental depletion.

Adhesion promoters are organic molecules that act as a tie layer between the polymer surface and adjacent material substrate. They typically change the surface chemistry making the two substrates chemically compatible. When added to the polymer formulation, it will take time and heat for the adhesion promoter to migrate from the bulk of the polymer to the interface. Therefore, the adhesive strength is dependent on processing conditions, and it will increase as a function of time after processing has been completed.

Fillers and foaming agents are primarily compounded for cost reduction but will also influence mechanical properties. Inorganic fillers, such as glass, talc, or calcium carbonate, are cheaper and mechanically harder than polymers. Filled thermoplastics are commonly used as structural members in PV frames due to high impact strength and low creep. In contrast, foaming agents react during processing to generate molecular gas, such as carbon dioxide, thereby increasing part porosity. This technique decreases costs and impact strength but increases mechanical creep. Foamed parts represent a larger structural risk for PV frames and must be thoroughly evaluated prior to use.

Processing agents can be described as modifiers and stabilizers. They are added to modify polymer properties in order to stabilize the chemistry against potential degradation endured during processing. Processing agents immediately relevant for this discussion are heat stabilizers, mold release agents, and viscosity modifiers. Heat stabilizers were previously discussed. Mold release agents are present in the thermoplastic formulation to allow for part ejection from the mold without warp. Finally, viscosity modifiers are included to improve flow of the polymer into the mold. They decrease frictional wear and ensure the molds are completely filled. These processing topics are covered in more detail in Chapter 4.

3.1.3 Other Classifications

Stabilizers and modifiers can be further classified as additive or reactive. Additives are small organic molecules subject to diffusion through the polymer. This mobility can benefit their efficacy. For instance, UV screeners must be at the polymer surface to adequately shield the chains from harmful radiation. However, as previously noted, these molecules can be depleted through evaporation or washing. A reduction in stabilizer concentrations can impact the polymer's lifetime and warranty. To avoid depletion, reactive additives are covalently bonded to the polymer chains. UV stabilizers are one of the most common reactive additives. Their environmental stability and resistance to migration make reactive additives a good fit for PV applications.

Polymers will not necessarily include all the aforementioned stabilizers and modifiers. A combination of compatibility with the polymer resin, design requirements, and cost targets influences the composition of the additive package. However, most polymers require additional additives to provide the desired mechanical, thermal, and weathering performance demanded by the PV industry.

3.2 The Effect of Additives on Polymeric Properties

Although additive packages are a proprietary part of the formulation, sometimes the major component is obvious from the polymer's grade. For instance, Rynite® is a trade name for polyethylene terephthalate sold by DuPont (Wilmington, Delaware). It is offered in various grades, including general purpose, toughened, and flame retardant (Table 3.2) [3]. General purpose grades are sold with varying glass fiber content imparting a range of impact strengths but a low flammability rating. Toughened Rynite contains reinforcement fillers to increase the Izod impact strength. Unfortunately, the additional filler decreases thermal conductivity. Flame retardant grades have high concentrations of flame retardants, improving the UL flammability rating. However, the addition of flame retardants lowers their impact strength. Therefore, optimizing one polymer behavior typically alters another. Care must be taken when choosing a commercial grade in order to ensure all critical performance criteria are met.

TABLE 3.2

Changes in Izod Impact Strength, Thermal Conductivity, UL Flammability, and Volume Resistivity of General Purpose Rynite, Toughened Rynite, and Flame Retardant Rynite

Different Commercial Grades of DuPont Rynite	Izod Impact Strength 296 K (J/m)	Thermal Conductivity (W/(m • K))	UL Flammability Rating	Volume Resistivity (Ω • cm)
General Purpose Rynite	117–69	0.33–0.29	HB	10^{15}
Toughened Rynite	235–133	0.26	HB	10^{15}–10^{13}
Flame Retardant Rynite	96–48	0.37–0.22	V-0	10^{15}

Source: Data from DuPont, 2010, DuPont Rynite® PET Technical Data Sheet, http://plastics.dupont.com.

3.3 Common Failure Mechanisms in Photovoltaic Packaging

Solarex was a successful U.S. PV manufacturer incorporated in 1973 and specializing in single crystalline silicon PV cells. In 1983, it merged with Amoco Solar Company, a subsidiary of Standard Oil, and by the end of the 1980s, Solarex had diversified into polycrystalline silicon. That segment of the business became profitable by 1994, catching the eye of British Petroleum [4]. In 1999, British Petroleum Solar merged with Solarex to become known as BP Solar.

As one of the oldest PV companies, they were in the unique position of having extensive field knowledge of terrestrial installations. Between 1994 and 2002, modules were collected from field applications and categorized by failure mode [5]. Although a root cause analysis was not provided for each observed failure, probable cause can be assigned to various polymeric components. For instance, the encapsulant can cause a number of different failures, including delamination, discoloration, corrosion, cell breakage, lead issues, arcing, overheated wires, and mechanical damage. In contrast, the junction box is typically suspected solely in electrical issues (Table 3.3).

These failure modes are caused by inadequate polymer properties that can be addressed by altering the polymer's formulation. For instance, coupling agents can be included in polymer formulations to enhance their adhesion to adjacent substrates and inhibit delamination. Polymer discoloration can be addressed through improved polymer stability with UV stabilizers and antioxidants, and corrosion can be inhibited with metal deactivators. The materials used by PV manufacturers today reflect decades of formulation optimization to prevent these failure modes.

TABLE 3.3

Common Failure Modes for Photovoltaic (PV) Modules with the Corresponding Packaging Component at Fault for the Failure

Common Failure Modes for Returned PV Modules between 1994 and 2002	Encapsulants	Frames	Junction Box	Backsheets
Corrosion	•	•	•	•
Cell or interconnect break	•			
Output lead problem	•			
Delamination	•			
Discoloration	•	•		•
Arcing	•		•	
Overheated wires, diodes, or terminal strip	•			•
Mechanical damage	•	•		•

3.4 Encapsulants

The National Aeronautics and Space Administration (NASA) launched its first satellite in 1958; however, by 1965 it launched more than 100 satellites per year. Each of these satellites was dependent on PV power with an intended module life span of 8 to 10 years. This expectation necessitated highly engineered polymeric packaging in order for the module to operate in harsh environments and remain maintenance free. Specifically, the packaging had to endure high irradiance (0.08 to 1.77 $Wm^{-2}nm^{-1}$ from 280 to 600 nm) and dramatic temperature swings (148 to 413 K) [6]. During this decade, polymer science was still in its infancy, and silicone encapsulants were the only polymers that met those requirements.

When PV modules were commercialized in the late-1970s, the encapsulant was the focus of multiple studies to find a material with reduced engineering requirements and lowered cost. Because the Earth's ozone layer filters most of the solar UV irradiance, the Sun's terrestrial intensity is three-quarters of the extraterrestrial intensity found in space (0.00 to 1.32 $Wm^{-2}nm^{-1}$ from 280 to 600 nm). Temperature differentials depend on the land geography and altitude, but for most regions, diurnal thermal differentials of 35 to 55 K are common. The lower irradiance and temperature differential decreased the durability requirements and broadened the polymers considered for this application.

TABLE 3.4

Material Specifications for Polymeric Encapsulants Used in Photovoltaic (PV) Module Packaging

Material Parameter	Specification
Optical	>98% optical transparency over 363 to 233 K and the visible spectrum
Thermal	T_g less than 233 K, thermal conduction 0.3 to 0.2 W/(m • K)
Mechanical	Low modulus ≤13.8 MPa at 298 K, no mechanical creep at 363 to 233 K, peel strength >1800 N/m
Electrical	High-volume resistivity 10^{16} to 10^{14} Ω • cm, high dielectric strength 10^5 to 10^4 V/mm
Flammability	Hot wire ignition (HWI) ≤ 4, high-current arc ignition (HAI) 3 to 2
Weathering/ transmission rates	No significant change in ΔYI, low oxygen transmission rate (OTR) 10^{-3} to 10^{-4} cm³/m²/day, low water vapor transmission rate (WVTR) 10^{-3} to 10^{-4} g/m²/day

In addition to weathering stability, encapsulants must fully adhere to adjacent components in order to provide the desired optical and mechanical performance. There must be good interfacial adhesion between each of these substrates to maximize optical transmission and eliminate entrapment of oxygen and water (Table 3.4). The glass and backsheet are two separate chemistries with different polarities. For instance, glass is typically polar, "water loving," while a fluorinated backsheet is nonpolar, "water hating." It is difficult to find a single polymer that adheres equally well to both substrates. Therefore, adhesion promoters are added to encapsulant formulations to improve adhesion and ensure a peel strength greater than 1800 newtons per meter (N/m). This is a minimum mechanical requirement, at both interfaces, for material selection.

The encapsulant must have the appropriate mechanical properties to prevent loading onto the embedded PV cells during thermal cycling. When there is good adhesion, mechanical stress is transferred from the encapsulant to the cells. If the encapsulant has a high elastic modulus, then during thermal expansion, the polymer will not readily yield. This increases stress at the interface, and it can result in PV cells cracking. For this reason, encapsulants must have a low elastic modulus but must not creep at temperatures between 233 and 363 K. If there is mechanical creep, the thickness will change during cycling, creating an uneven stress distribution also leading to cell fracture. Elastomers and ionomers are the most commonly used encapsulants, because they meet these mechanical requirements.

Finally, the encapsulant must be a good electrical and environmental insulator to prevent electrical drift and permeant diffusion. Electrical requirements are a minimum of 10^{14} ohms-centimeter (Ω • cm) for volume resistivity and 10^4 to 10^5 volts per millimeter (V/mm) for dielectric strength. Any drifts below these values can lead to leakage current and arcing between cells. In addition, oxygen and water vapor transmission rates must be on the order of 10^{-3} to 10^{-4} cm³/m²/day and 10^{-3} to 10^{-4} g/m²/day, respectively. This ensures the electronic components are appropriately protected to prevent corrosion during use.

3.4.1 Polysiloxane

U.S. companies and researchers were instrumental in the successful com-mercialization of polysiloxane chemistry. But German chemist Friedrich Wöhler and English chemist Frederick Kipping were academic researchers credited as the fathers of polysiloxane chemistry. Despite their pioneering developmental work, neither Wöhler nor Kipping recognized the commer-cial importance of their molecules. It was Eugene Rochow, a chemist for General Electric (GE) (Fairfield, Connecticut), who patented polydimethyl-siloxane in 1941 as a first step toward commercialization. Simultaneously, Dow Chemical (Midland, Michigan) had also begun to work on the process engineering required for mass production of polysiloxanes. Soon the growth in the business was large enough for Dow Chemical to partner with its sili-con supplier, Corning Glass Works (Corning, New York). Together the two companies formed the subsidiary Dow Corning, which pioneered a signifi-cant amount of formulation development in the late 1950s and early 1960s. This development work and resultant product offering occurred during the same time the first prototype PV modules were produced. Therefore, both GE and Dow Corning products were used for PV manufacturing and are commonly referred to in the PV literature [7]. However, because GE sold its silicone division in 2006, some of these historical products are now offered by Momentive (Albany, New York).

Polysiloxanes, also known as silicones, are composed of alternating silicon atoms bonded to oxygen forming a long polymer chain. One silicon–oxygen (Si-O) bond is present in each repeat unit. Polysiloxanes are differentiated by the groups of atoms bonded to the silicon atom. The substituent names are inserted in between the prefix *poly-* and the word *siloxane* in the nomen-clature. For instance, polydimethylsiloxane indicates two methyl groups are attached to each silicon atom (Figure 3.1[I]).

The substituents influence the polymer's refractive index. When two methyl groups are included, then the refractive index (n_D) is 1.40. If one methyl sub-stituent is replaced with a phenyl group (Figure 3.1[II]), the polymer is known as polymethylphenylsiloxane, and the refractive index increases to 1.53 [8]. Design engineers can request an intermediate refractive index by designat-ing the desired phenyl content included in the silicone chain.

The strong silicon–oxygen bond in silicones gives them high UV and ther-mal stability. Because of this inherent stability, silicones do not require the glass superstrate to shield UV light from the underlying polymer chains in order to last the 25 to 30 years required by PV manufacturers. In addition, they have a service life of 40 years at 363 K and 10 to 20 years at 394 K based solely on oxidative degradation studies [9].

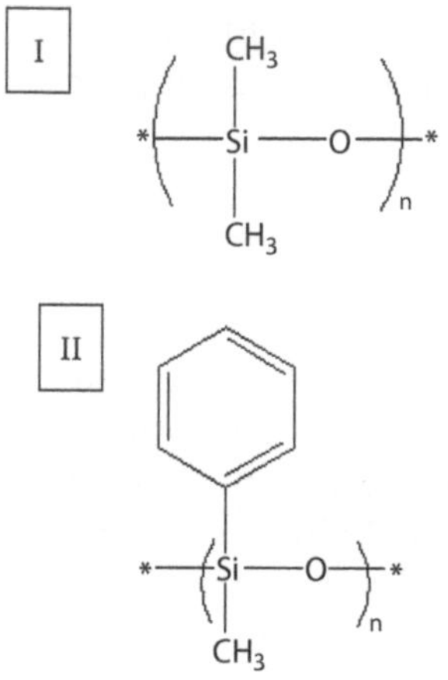

FIGURE 3.1
Chemical structure of (I) polydimethylsiloxane and (II) polymethylphenylsiloxane.

Due to their higher costs, silicones are a logical choice for small-quantity applications with extreme UV and temperature requirements [10]. For instance, they have been used in terrestrial concentrated photovoltaic (CPV) devices for years. CPV uses mirrors and lenses to concentrate the Sun's rays. Depending on the module design, the Sun's rays can be magnified to 2 to 500× regular irradiance. Increased irradiance has a significant impact on the device's temperature. As an example, a 500× increase in UV irradiance heats the encapsulant to 423 to 473 K [11].

Commercial formulations are a combination of silicone polymers, cross-linking agents, fillers, and processing agents. The most basic component of the formulation is the silicone polymers that constitute the base. In order to form an elastomer, cross-linking agents are included in the formulation. These small molecules initiate a chemical reaction to link the polymer chains into a three-dimensional network. Extending fillers are added to the formulation in 70 to 80 wt% concentrations to reduce the cost of the polymer and extend the formulation weight. These high filler loadings alter the polymers' inherent properties. Silica and aluminum trihydrate are commonly added to silicones to increase mechanical strength and impart flame resistance. High filler content can also increase the formulation's resistance to flow, thereby inhibiting processing. Filler content should be optimized to minimize processing issues and maximize mechanical properties. However, silicone oil can also be added to offset the thixotropic effects of fillers and improve flow.

Room-temperature vulcanates (RTVs) are a one-part chemistry purchased from the manufacturer in an airtight drum or canister, and they require an environmental agent to cross-link. One of the most popular is a condensation cure RTV. For this chemistry, acetoxy groups on the methyltriacetoxysilane cross-linking agent react with humidity to form alcohol groups (Figure 3.2). In the presence of a tin catalyst (e.g., dibutyltin laurate), the alcohol group

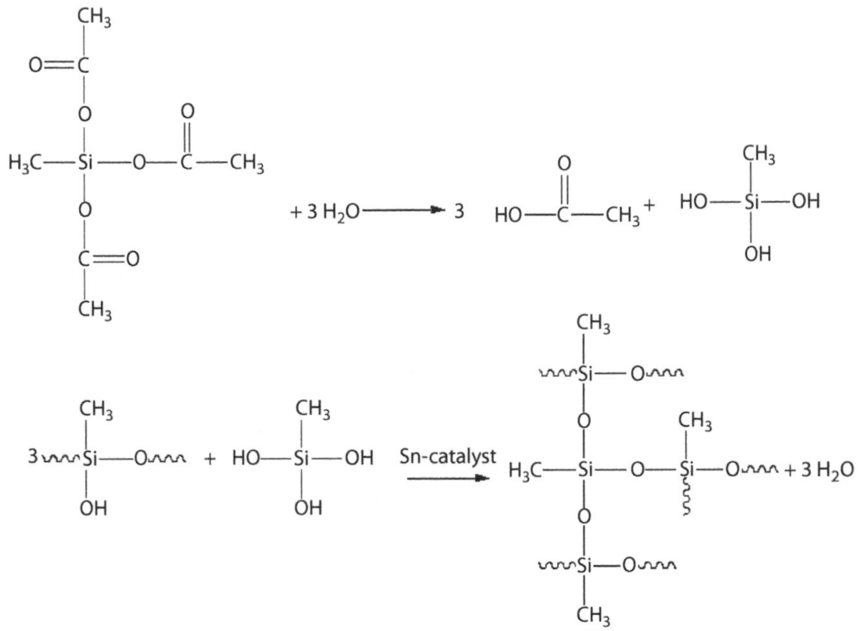

FIGURE 3.2
Example of a one-part silicone reaction.

reacts with the base to form a cross-linked elastomer. Carboxylic acids (i.e., acetic acids) are released in this example, but small molecule by-products, including alcohols, amines or amides, can also be produced in variations of these reactions. The specific by-product depends on the cross-linking agent and is commonly specified in the manufacturer's literature.

One-part chemistries are prone to some common failure mechanisms. In the late 1970s, NASA experienced corrosion of potted electronic components due to the presence of corrosive by-products. The polymer market responded with a new chemistry called noncorrosive, one-part silicones for electronic applications. PV manufacturers should look for this product designation when choosing chemistries for electronic pottants. Regardless, all condensation RTVs are inhibited by a high, localized concentration of by-products, making them an unacceptable selection for use in confined spaces. Also, the presence of tin catalysts has been blamed for premature encapsulant yellowing.

One-part chemistry requires environmental permeants to diffuse through the thickness before the base can completely cure. Typically, a skin of cured polymer will form on the surface slowing the penetration of the coreactant through the bulk. Therefore, complete cure can take days in low-humidity environments. However, these silicones are less susceptible to manufacturing errors because the two components are not mixed by the manufacturer.

Two-part silicones include a base chemistry and catalyst, both supplied by the polymer manufacturer. The two components must be precisely mixed at the specified ratio to ensure proper cure. The most common two-part cure chemistry is described as addition reactions (Figure 3.3). The base, a polysiloxane with pendant vinyl groups, reacts with a polysiloxane reagent containing a silicon–hydride (Si-H) bond, in the presence of a platinum catalyst. As the description implies, cross-links are added across the double bonds of the vinyl groups on the polysiloxane chains. These reactions do not emit a by-product, but the catalyst remains in the elastomer. Adjacent butyl and chlorinated rubbers, sulfur-containing materials, and organo-tin molecules inhibit this cure reaction.

Most PV applications use two-part silicone chemistries because they cure quickly, shrink less, and absorb less moisture. A faster cure time shortens the manufacturing cycle, consequently increasing manufacturing output. Two-part chemistries begin to immediately cure throughout the bulk once the catalyst is blended into the silicone base. In addition to the faster manifestation of cure properties, two-part silicones shrink less (0.10 to 0.15% versus 0.4 to 0.6%) because there are no small molecules evolved during cure. Most important to PV packaging, a two-part silicone has a lower water absorption (0.1 to 0.12% for two-part versus 0.2 to 0.4% for one-part) and does not evolve moisture during the cure reaction [12].

3.4.2 Polyvinyl Acetate and Polyethylene

Chain length and architecture both influence crystallinity. Low molecular weights and structural regularity decrease the spatial proximity between chains and therefore increase their propensity to form crystalline domains. Crystallinity can be disrupted with increased molecular weight and bulky

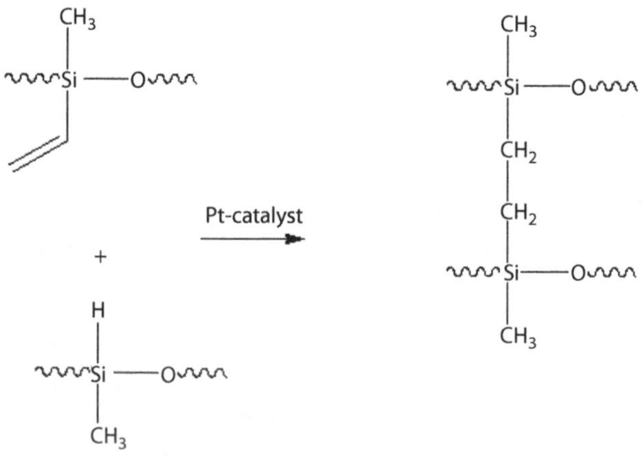

FIGURE 3.3
Example of a two-part silicone reaction.

branches along the chain's backbone. The two relevant polymers that demonstrate this relationship are polyvinyl acetate and polyethylene.

Polyvinyl acetates are constructed of a hydrocarbon chain with a pendant
acetate ester (Figure 3.4). Their percent crystallinity can range from 0.701% to
0.587% for 2236 g/mol to 16,856 g/mol, respectively [13]. Polyvinyl acetates
are usually highly branched and completely amorphous at higher molecular
weights.

The molecular architectures of polyethylene dictate the polymer's density.
Polyethylene has a hydrocarbon backbone and pendant hydrocarbon side
chains (Figure 3.5). The pendant chains, termed branches, occur in random
frequency. Their branch length influences the polymer's density. High density polyethylene is characterized by a few short chains forming branching

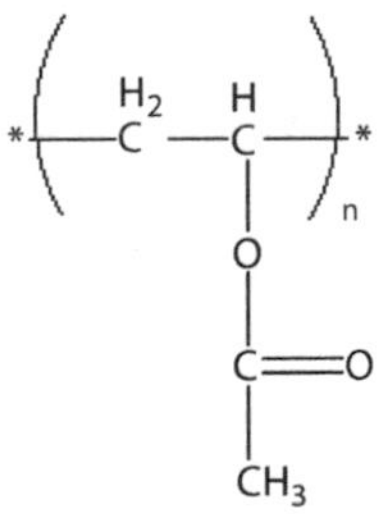

FIGURE 3.4
Chemical structure of polyvinyl acetate.

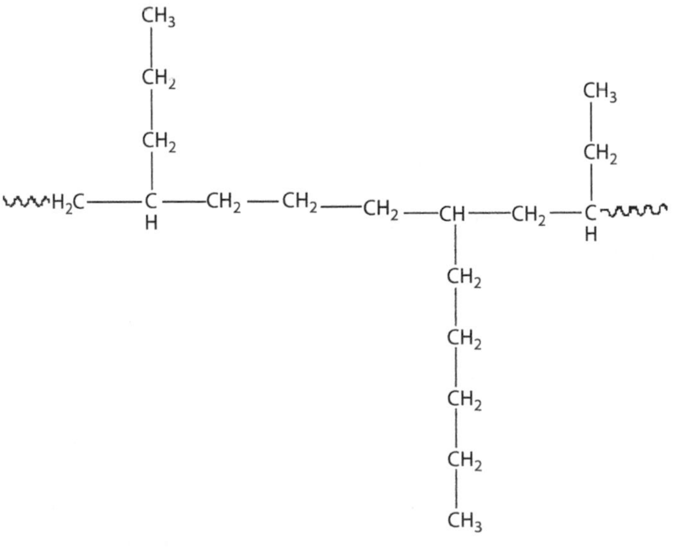

FIGURE 3.5
Chemical structure of a section of polyethylene.

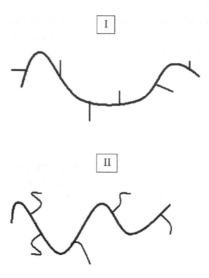

FIGURE 3.6
General depictions of a (I) high-density polyethylene (HDPE) chain and a (II) low-density polyethylene (LDPE) chain.

points randomly located along a long hydrocarbon backbone, Figure 3.6. This allows for efficient chain packing into crystalline domains composing a large portion of the polymer's matrix (70% to 90%) and, subsequently, a higher polymer density (0.94 to 0.96 g/cm^3). Low-density polyethylene is composed of long branches limiting the physical proximity of adjacent chains. The crystallinity decreases (40% to 60%), resulting in a lower density (0.91 to 0.93 g/cm^3) [29].

3.4.3 Ethylene Vinyl Acetate Copolymer

Ethylene vinyl acetate (EVA) is a copolymer composed of ethylene and vinyl acetate monomers. Chemically linking the two monomers creates a blend of the corresponding homopolymer's properties. For instance, linear chains of polyethylene crystallize to form a brittle plastic. By copolymerizing with vinyl acetate, the polymer's crystallinity is decreased, and the mechanical properties can be tailored to the application's requirements (Table 3.5). EVA has an intermediate glass transition, melt temperature, and elongation to break between the two homopolymers.

Typical commercial EVA formulations contain 2% to 40% vinyl acetate. At 11% vinyl acetate concentration, EVA will be a rigid, solid at room temperature and is used in carpet backings and in hot melt adhesives. At higher vinyl acetate concentrations, EVA is flexible at room temperature and is used in food packaging, shrink-wrap films, and chemical drum liners. The most common weight percent for PV applications is 33% vinyl acetate making it a flexible substitute.

TABLE 3.5

Glass Transition, Melt Temperature, and Elongation to Break of Polyvinyl Acetate, Ethylene Vinyl Acetate Copolymer, and Low-Density Polyethylene

Polymer Properties	Polyvinyl Acetate	Ethylene Vinyl Acetate Copolymer	Low-Density Polyethylene
Glass transition (T_g, K)	304–297	235–231	170–140
Melt temperature (T_m, K)	448	379–318	388–378
Elongation to break (%)	20–10	850–675	800–100

Sources: Data from Mark, J.E., 1999, *Polymer Data Handbook*, Oxford: Oxford University Press; Mark, H.F., 1985, *Encyclopedia of Polymer Science and Engineering*, 2nd Vol., 15th Ed., New York: Wiley.

EVA is sold by a number of manufacturers, because it is the most popular type of polyethylene copolymer. The DuPont Corporation first launched EVA in the 1960s, and it remain one of the largest producers (sold under the trade name Elvax®). There are other sources of EVA, principally Escorene™ sold by ExxonMobil (Houston, Texas) and Ultrathene® sold by Equistar Chemicals (Houston, Texas). All these commercial formulations include a proprietary combination of processing agents, UV stabilizers, and antioxidants.

EVA is a thermoplastic; therefore, when heated above its melt temperature, it will flow. During the late 1970s, when researchers were searching for encapsulants to be used in terrestrial PV modules, most commercial formulations flowed into a new shape at 343 K. Because this was within the expected service temperature of terrestrial PV modules, it was not an acceptable material behavior. Furthermore, the polymer's crystallinity decreased its optical clarity. Therefore, formulators started to place peroxide additives in EVA formulations. During processing, the peroxides reacted with the polymer to create a three-dimensional matrix. Once cross-linked, there is no concern that the encapsulant will reform crystallites after processing or flow during service.

Unfortunately, chemical cross-linking agents were a source of chemical instability during reliability testing. Elvax 150 is a grade of EVA that included Lupersol 101, 2,5-dimethyl-2,5-*bis*-(t-butyl peroxy) hexane, as a cross-linking agent. Under low UV exposure and thermal aging, Elvax 150 showed signs of yellowing. More concerning, test modules bubbled when thermally cycled. In both cases, the degradation mechanism was tied to the chemical instability of the unconsumed Lupersol 101. Specifically, Lupersol 101 was decomposing into ethylene and ethane gas during aging. Today, Lupersol 101 has been removed from most commercial formulations and replaced with a more stable, proprietary peroxide.

EVA formulations also contain UV stabilizers to extend their outdoor service life. Pern and coworkers reported on the chemical deformulation of darkened encapsulants removed from the field. They found a correlation between the absence of UV absorbers and coloration. Specifically, decreased concentrations in Cyassorb UV 531, a UV absorber, and Tinuvin 770, a UV stabilizer, were associated with EVA's yellow coloration. The yellow coloration will darken into a brown coloration with complete stabilizer depletion [14,15]. The coloration has been the subject of numerous studies and is believed to be the result of chromophore formation in EVA chains. Specifically, the coloration is caused by a polyene, described as conjugated alkene (C=C) bonds along the polymeric backbone [16].

The rate of color formation and its effect have been reported with varying degrees of severity on module performance. For instance, some have reported no change in power despite encapsulant yellowing. Conversely, others report 40% power loss due to dark brown discoloration within 5 years of field exposure. Typically, those locations with the highest temperatures and solar irradiance were the first locations to exhibit discoloration; however, the degradation times ranged from 3 to 12 years [17].

The type of superstrate glass used in the module can significantly influence EVA's stability. Various glass compositions will have different UV cutoff ranges. For instance, more UV-A (320 to 400 nm, 51×) and UV-B (280 to 320 nm, 126×) is transmitted through nonceriated glass versus cerium-containing (also know as ceriated) glass (Figure 3.7). Ceriated glass is a necessity when using EVA as an encapsulant. More specifically, at 60°C (333 K), 60% Relative Humidity (RH), and 2.5 × UV suns, EVA's lap shear strength starts to decrease from 10 MPa to 3 MPa after 100 hours behind nonceriated glass. When weathered under the same conditions behind ceriated glass, this decrease occurs at 1000 hours. It is concluded that the ceriated glass significantly reduced UV-B

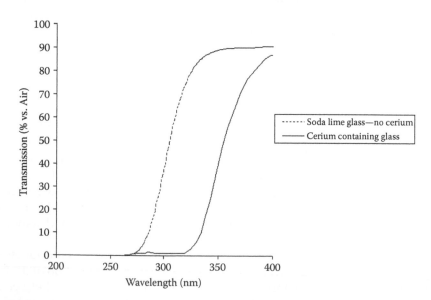

FIGURE 3.7
Ultraviolet-visible (UV-Vis) transmission spectra of ceriated and nonceriated glass.

irradiance, improving adhesive strength. However, the effect of UV-A and humidity were not separately documented in these experiments [18].

In order to effectively insulate and protect the encapsulated electrical components, the penetration of moisture from the surroundings must be minimized. When insufficiently protected from moisture ingress, the corrosion of electronics has been attributed to localized acid formation in the EVA. EVA decomposes via a hydrolysis reaction to form acetic acid. Acid generation has been reported during both thermal aging and weathering experiments, and it is correlated with the encapsulant yellowing [18]. For this reason, the water vapor transmission rate (WVTR) of the superstrate and backsheet must be minimized.

3.4.4 Polyvinyl Butyral

Polyvinyl butyral (PVB) is a thermoplastic with a variable chemical structure. This variability is due to unreacted monomeric units in the polymer chain (Figure 3.8). It is synthesized from polyvinyl alcohol and forms polyvinyl acetate as a reaction intermediate. Commercial formulations typically have 11 to 20 wt% unreacted polyvinyl alcohol and 0 to 2.5 wt% polyvinyl acetate [13].

Safety glasses are the principal commercial use of PVB. However, when used for safety glasses, the formulations contain a plasticizer to decrease brittleness and increase shatter resistance. Its optical clarity made PVB an obvious choice for qualification testing as a PV encapsulant.

When first tested in the late 1970s, it was PVB's commercial formulation and not the inherent polymer properties which led to its disqualification. An increase in plasticizer content decreases the volume resistivity from 10^{16} $\Omega \bullet cm$, unplasticized, to 10^{11} $\Omega \bullet cm$, plasticized, and increases the module leakage current from 2 µA to 25 µA. When tested from room temperature to 60°C (333 K), the leak current increases by another order of magnitude from 25 µA to 250 µA [19]. Leakage current decreases the overall module power; therefore, PVB was deemed an unsuitable encapsulant.

Of the total amount of PVB produced in 1994, 97% contained plasticizer; therefore, PVB has not recently been reconsidered as a possible PV

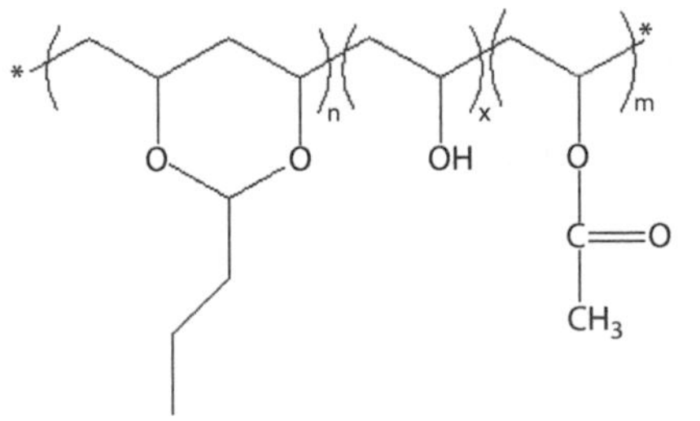

FIGURE 3.8
Chemical structure of polyvinyl butyral (PVB).

encapsulant [13]. However, the lesson from this study is that new materials considered for encapsulation must not contain plasticizer additives.

In 1983, Lewis and Megerle noted returned PV modules with PVB encapsulants exhibited browning above lead contacts. When the chemical structure of the polymer was examined, abnormal concentrations of carbonyl and oxygen signals were found. This suggested the polymer had undergone oxidative degradation. The mechanism could be duplicated when PVB was thermally aged in the presence of vanadium oxide, antimony oxide, or mixtures of copper and nickel oxides. These are the same components used in the glass frit of PV cell grid lines. Therefore, it was postulated these metal centers catalyzed the oxidative degradation of the polymer [19]. In this example, the discoloration did not result in power loss because the discoloration was not above the PV cells. However, it is probable the degradation could have affected a larger surface area if it had been left in the field longer. This finding further increased the industry's disinterest in this material as an encapsulant.

3.4.5 Ionomers

Immediately following World War II, researchers were interested in imparting the unique properties of elastomers to thermoplastic polymers. Originally, researchers attempted ionization, using electron beam radiation, and chemical techniques, such as peroxide treatments, to cross-link thermoplastic polymeric chains. However, it was soon discovered that a blend of thermoplastic and elastomeric properties could be achieved with block copolymers.

A thermoplastic elastomer (TPE) is a block copolymer that organizes into soft, amorphous and hard, crystalline domains. The hard domains act as physical cross-links to prevent the soft segments from flowing. Unlike chemical cross-links, these physical cross-links are reversible and can be removed with heat. Specifically, the hard, crystalline domains dissolve, and the polymer flows at temperatures above the melt temperature. This makes TPEs suitable for service temperatures below the melt temperature of the hard phase.

Ionomers are a specific type of TPE. Ionomers are block copolymers containing polar and nonpolar segments. A neutralized block copolymer has no ionic charges, but it will segregate to form hard- and soft-phase morphology. When the polar segments are acidic, those segments ionically bond to a cation to neutralize their charge. The ionic segments sequester into ionic domains and behave as additional physical cross-links.

Manufacturers sell these polymers based on the percentage of acidity, also called ionization, and type of counterion. There are a number of counterions available, but zinc, sodium, and magnesium are the most common. How each of these affects the material properties is best represented in the data sheets from the manufacturer.

The first commercial success for these polymers occurred in the early 1950s when DuPont marketed Hypalon®, a chlorosulfonated polyethylene. It was not until 1964 that the word *ionomer* was coined by R.W. Rees to describe Surlyn's morphology. His pioneering work developed the Suryln product offering available to today's consumer.

3.4.5.1 Surlyn®

Polyethylene-*b*-polymethacrylic acid salt-*b*-polymethylacrylate is an iono-mer sold by DuPont under the commercial trade name Surlyn. Currently, there are over 30 grades of Surlyn available. The chemistry of those grades is based on work performed in the early 1960s by R.W. Rees, while employed at DuPont. Holden and coauthors recount this work in *Thermoplastic Elastomers* [20]. Particularly relevant to this audience were their observa-tions of changes in mechanical strength and moisture absorption with chemical structure.

Increases in vinyl acetate content in the block copolymer will break up the crystallinity in the polyethylene regions and alter mechanical properties. For instance, the secant modulus for EVA decreases from 75.9 MPa for 9% vinyl acetate content to 18.6 MPa for 28% vinyl acetate content [13].

Mechanical behavior is also dependent on the percent ionization in the polar segments. The tensile strength increases with additional ionization and reaches a plateau at 77%, irrespective of the molar percent of vinyl ace-tate present on the polymer chain [20]. Maximum tensile strength values are between 25 to 50 MPa for 1.7 to 5.9 molar percent vinyl acetate, respectively.

The moisture uptake properties are significantly influenced by the counte-rion present. Initially, the moisture behavior of these polymers was crudely measured by observing changes in weight gain when the polymers were placed in boiling water. During these initial developmental studies, Rees synthesized polymers with sodium, potassium, lithium, magnesium, zinc, strontium, and lead. Strontium (0.16 wt%) and lead (0.13 wt%) counterions had a similar uptake capacity as the neutralized polymer (0.13 wt%). A sodium ionomer (2.25 wt%) results in 10x moisture gain relative to a zinc ionomer (0.22 wt%) [20]. These two counterions bracketed the extremes of the observed behavior and represent the commercial limits offered by DuPont. Sodium and zinc ionomers have been offered in DuPont's prod-uct line since 1965 and are currently sold under the 8000 and 9000 product codes, respectively.

The weathering characteristics are dependent on both the stabilizer pack-age and the cation ion chemistry. Transparent Surlyn with a combination of antioxidants and UV absorbers will retain the majority of its impact and ten-sile properties after 5 years of Arizona outdoor exposure. Furthermore, grade 9000 Surlyns will have a higher retention of mechanical properties than 8000-grade Surlyns, when stabilized with the same concentration of antioxidants, UV absorbers, and hindered amine light stabilizers. In addition, most com-mercial grades contain a few parts per million (2 to 10 ppm) of Monastral blue in order to offset any yellow coloration due to polymeric weathering [21].

3.5 Frames

In 1983, the researchers tasked with finding polymers for terrestrial PV applications admitted plastics were a natural substitute for metal frames. They also stated in their reports that "an optimum candidate has not yet been identified" [19]. Today, not much has changed. Most commercial frames are composed of anodized aluminum. There are a few exceptions; Sun Power's (San Jose, California) T60 and Concentración Solar La Mancha S.L.'s (Ciudad Real, Spain) concentrated photovoltaic modules both have a thermoplastic frame. As the PV industry continues to strive for cost-saving measures, more modules will likely integrate polymeric frames.

Mechanical properties are the main functional requirements for this application (Table 3.6). The polymer must have a tensile stress at break higher than 68.9 MPa and a flexural modulus above 3450 MPa. These are guidelines, but the absolute limits are dependent on the exact design requirements. Regardless of the specific mechanical values, the polymer must not exhibit mechanical creep below 90°C (363 K). To exhibit these mechanical specifications and endurance, most thermoplastics require reinforcement fillers.

To date, the two most common engineering thermoplastics for frames are glass-filled polyesters, such as Rynite sold by DuPont, and acrylonitrile-styrene-acrylate (ASA) copolymer, such as Luran S® sold by Badische Anilin-und Soda-Fabrik (BASF) (Florham Park, New Jersey). Both types of polymers have been marketed for outdoor applications, such as luggage racks for automobiles and lawn furniture. Polymer manufacturers primarily base these recommendations on color shift resistance and mechanical property degradation after 3 to 5 years of outdoor testing. Therefore, the outdoor durability of these polymers for multiple decades as structural components requires careful scrutiny by the PV manufacturer.

Unlike anodized aluminum, polymers are not susceptible to corrosion, but they do undergo environmental degradation. As previously noted,

TABLE 3.6

Material Specifications for Polymeric Frames Used for Photovoltaic (PV) Balance of System (BOS) Components

Material Parameter	Specification
Mechanical	No mechanical creep at 363 K Tensile stress at break >68.9 MPa High flexural modulus >3450 MPa
Flammability	V-0 rating or better
Weathering	f1 UL rating, no color shift deemed unacceptable by the consumer and no drop in mechanical properties below the above thresholds, no degradation that significantly reduces the mechanical properties or seal integrity

polyesters are susceptible to hydrolytic degradation, and this makes high-humidity installations a concern. For ASA, little is known about the degradation mechanisms, but gloss retention is generally poor. Their questionable environmental stability has precluded reinforced thermoplastics from widespread use for PV frames.

3.6 Junction Boxes

Polymers can be used for both junction box enclosures and pottants, but each application has different certification and material requirements. Of the two components, the enclosure has the most stringent certification requirements.

An enclosure is defined as an open container that is not air- or watertight. It houses the encapsulated electrical components providing structural support and a partial environmental barrier. It must have a high V rating under UL 94, exhibiting limited combustion and flame spread characteristics (Table 3.7). The relative temperature index (RTI) of the polymer must be greater than or equal to 363 K. This requires the polymer to exhibit high transition temperatures. Therefore, reinforced thermoplastics are the most widely used classification for this application.

For a pottant, the flammability and environmental stability are greatly reduced because it is a secondary environmental barrier. The focus is on the polymer's electrical properties. The polymer must have a high dielectric constant and dielectric strength to provide insulation for the electrical components (Table 3.8). The two most widely used pottants are silicones and epoxies. The RTV product line sold by Momentive and Sylgard® sold by Dow Corning are some commercial pottants that have been used with previous success. Huntsman's Epocast® (Los Angeles, California) and Emerson & Cuming's Stycast® (Billerica, Massachusetts) are popular epoxy pottants for PV applications.

TABLE 3.7

Material Specifications for Polymeric Junction Box Enclosures Used for Balance of System (BOS) Components

Material Parameter	Specification
Thermal	Relative temperature index (RTI) equal or above 363 K
Mechanical	Impact resistance >22.6 N • m
Electrical	Comparative tracking index (CTI) ≤2
Flammability	Flammability rating of 5-VA, hot wire ignition (HWI) ≤4, high-current arc ignition (HAI) 3-2
Weathering	f1 UL rating

TABLE 3.8

Material Specifications for Polymeric Junction Boxes Pottants Used for Photovoltaic (PV) Balance of Systems (BOS) Components

Material Parameter	Specification
Thermal	High thermal conductivity <0.4 W/(m • K), low coefficient of thermal expansion
Electrical	Volume resistivity 10^{16} to 10^{14} Ω • cm, comparative tracking index (CTI) ≤2
Flammability	Flammability rating of HB, hot wire ignition (HWI) = 1 for HB materials, high-current arc ignition (HAI) = 1 for HB materials
Transmission rates	High water vapor transmission rate (WVTR) <10^{-1} g/m²/day

3.7 Backsheets

Backsheets are a primary environmental barrier to the encapsulated PV cell. Therefore, backsheets must be virtually impermeable to environmental ingress to insure electrical cell interconnects are protected from corrosion. To avoid this mechanism, backsheets have traditionally been composed of glass. However, even with these structures, moisture and oxygen could ingress through the laminate's edges. Aluminum encapsulated in EVA and sandwiched between two pieces of glass will corrode within 700 hours of 85°C (358 K)/85% RH exposure, also known as damp heat. Acid formation from moisture catalyzed EVA degradation has been blamed for corrosion of the encapsulated aluminum [22]. In addition, glass is mechanically fragile and can easily crack and break when impacted during installation and storms.

Currently, the industrial norm is to use cheap, lightweight polymeric backsheets. In order to be a successful candidate, the polymer must exhibit low moisture ingress, high mechanical strength, high electrical resistance, minimal flammability, and high UV stability (Table 3.9).

There has been considerable controversy on the appropriate requirements for water vapor transmission rates (WVTRs) of backsheets. Some researchers advocate a high transmission rate, effectively making the system breathable. Although moisture can easily ingress into the module at night, it can also egress and evaporate when temperatures increase the following day. This prevents encapsulants from experiencing degradation mechanisms accelerated with the synergic effects of light, moisture, and heat. Alternatively, the counterargument is a low WVTR will limit the concentration of water in the module. However, these smaller amounts of ingressed water will be trapped inside the module and available to participate in degradation processes [23].

TABLE 3.9

Material Specifications for Polymeric Backsheeets Used in Photovoltaic (PV) Module Packaging

Material Parameter	Specification
Optical	Solar reflectance >69%
Thermal	Thermal conductivity 0.2–0.1 W/(m • K)
Mechanical	Peel strength >1800 N/m
Electrical	High volume resistivity 10^{14}–10^{16} Ω • cm, high dielectric strength 1.18 • 10^5 V/mm
Flammability	V-0 rating or better, high burst pressure rating >0.35 MPa at 573 K
Weathering/ transmission rates	Color shift (ΔYI) <0.75, low oxygen transmission rate (OTR) 10^{-3}–10^{-4} cm³/m²/day, low water vapor transmission rate (WVTR) 10^{-3}–10^{-4} g/m²/day

Despite the advantages of both approaches, most manufacturers advocate a lower WVTR to minimize ingress [24].

Historically, modules suffer backsheet delamination after 1000 hours of damp heat (85°C (358 K)/85% RH). To minimize this failure mode, the backsheet must be well adhered to the adjacent encapsulant. Various coupling agents have been added to backsheets to improve adhesion. For instance, fluorinated backsheets are adhered to EVA encapsulants with a dry acrylic adhesive that is thermally activated during processing [26]. Delamination remains a widely discussed failure mechanism that must be investigated for each candidate backsheet.

The backsheet is subjected to a number of certification requirements. Most PV module certifications require a minimum V-0 flammability rating. The lower flame ratings of the encapsulant necessitate the backsheet have the required mechanical strength to confine and separate the encapsulant from impinging flames. Therefore, the backsheet must have high mechanical strength and burst pressure at elevated temperatures (Table 3.9).

The backsheet must also have the appropriate coloration to remove heat from the PV cells. In the late 1970s, a number of PV modules were constructed from black backsheets. The black backsheet absorbed solar irradiance, increasing the module's operating temperature. White backsheets have become the new industrial standard with a required solar reflectance of 69% or greater. In this color, the backsheet reflects infrared light and decreases device temperature. However, white polymers are more susceptible to color shifts during weathering. Therefore, there is a color shift specification of less than 0.75 unit change in yellowness index (ΔYI) over the 25- to 30-year lifetime of the PV module.

3.7.1 Fluorinated Polyolefins

The chemical structure of fluorinated polyolefins gives them the highly desirable properties of high thermal stability and low flammability. The small atomic size of the fluorine atom allows for efficient packing of the polymer chains into crystalline domains. With specialized processing conditions, up to 50% of the polymer can be composed of crystals. Fluorinated polyolefins typically melt between 463 and 600 K, depending on their chemical structure (463 K for Tedlar® and 600 K for Teflon®) [13]. Therefore, their suggested service temperature has a lower limit of 73 K and an upper limit of 60 to 70 K below the polymer's melt temperature. In addition, fluorinated polyolefins do not require expensive flame retardants. The chemical structure of the fluoropolymers means it cannot support combustion. The carbon–fluorine (108 kcal/mol) bond is stronger than other carbon–halide bonds, including carbon–bromine (70 kcal/mol). Therefore, the fluorine on the polymer chain is a form of reactive flame retardant. Due to these attributes, fluorinated polyolefins are thermoplastics used for highly engineered applications. These polymers are typically used in low-volume, high-value nonstick applications, such as wheel bearings, cookware, and medical devices.

In addition to these characteristics, fluorinated polymers exhibit low transmission rates and high resistance to humidity. Polyvinyl fluoride (PVF) (Figure 3.9) is the most common backsheet because it does not embrittle during damp heat exposure. Crystalline domains also decrease permeant transmission rates when randomly formed in the polymer's matrix. Due to the effect of processing on percent crystallinity, moisture and oxygen transmission rates vary widely based on commercial formulation and the manufacturer's processing conditions. As an example, polyvinyl fluoride's WVTR can range from $0.7 \cdot 10^{-4}$ to $1.54 \cdot 10^{-3}$ g/m²/day based on the manufacturer's grade [25]. Even though they improve transmission properties, the crystalline domains in PVF result in dimensional instability, and they will cause the polymer to shrink during damp heat exposure. Regardless, the other highly desirable properties make fluorinated polymers 3.5 times more expensive than other substitutes, such as polyesters [26].

3.7.2 Laminate Structures

To reduce cost but maintain the desired barrier properties, PV manufacturers commonly use laminate structures. DuPont's TPT™ laminates are a polyester, Mylar® A, sandwiched between two layers of polyvinyl fluoride, Tedlar. The fluorinated polymers provide superior environmental resistance, while the polyester layer imparts mechanical strength.

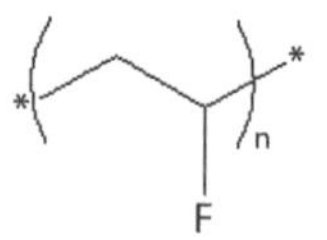

FIGURE 3.9
Chemical structure of polyvinyl fluoride (PVF).

To further reduce costs, polyester backsheets have been developed to completely eliminate expensive fluorine chemistry. Historically, polyesters have been avoided due to poor moisture stability. Specifically, polyethylene terephthalate (PET) shows rapid embrittlement after aging in 85°C (358 K) and 85% RH. The mechanical properties often deteriorate until the sample is so brittle the peel strength cannot be accurately measured. To reduce susceptibility to hydrolytic degradation, polyesters are either metallized, with silicon oxide (SiOx), or laminated with metal foil, such as aluminum. Both methods significantly reduce the WVTR, typically by at least an order of magnitude below that of the unaltered polymer [27]. Decreased permeability increases the laminate's resistance to damp heat degradation. For instance, silicon oxide–metallized PET demonstrates a modest drop in peel strength after 2000 hours of damp heat in comparison to unmodified PET, which cannot be measured due to embrittlement.

The adhesive strength between EVA and PET is typically 10 times higher than the adhesive strength between these various laminate layers [22]. During damp heat exposure, the laminated backsheet layers have been known to separate. If the polymer–metal interface separates, the metal foil can create an electrical short to the aluminum frame [28]. Therefore, metal foil laminates have fallen out of consideration for most PV manufacturers.

In effect, improving the homopolymer's UV and hydrolysis resistance has become a larger industrial effort. DuPont has developed new proprietary PET formulations. These new laminates are a combination of white, high-dielectric polyester, known as Teijin®, laminated to a UV and hydrolysis-stabilized polyester, Melinex®. Experimental data regarding this product's performance are primarily limited to the manufacturer's literature.

References

1. Ezrin, Myer. 1996. *Plastics Failure Guide: Cause and Prevention.* New York: Nahser.
2. Ackart, W.B.; Camp, R.L.; Wheelwright, W.L.; Byck, J.S. 2004. Antimicrobial Polymers. *Journal of Biomedical Materials Research* 9: 55–68.
3. DuPont. 2010. DuPont Rynite® PET Technical Data Sheet. http://plastics.dupont.com/
4. BP Solar. 2010. History. www.bp.com
5. Wohlgemuth, J.H. 2003. Long Term Photovoltaic Module Reliability. NCPV and Solar Program Review Meeting NREL/CD-520-33586 179–182.
6. Gilman, J.W.; Schlitzer, D.S.; Lichtenhan, J.D. 1996. Low Earth Orbit Resistant Siloxane Copolymers *Journal of Applied Polymer Science* 60: 591.
7. Chopra, K.; Das, S. 1983 *Thin Film Solar Cells.* New York: Plenum Press.
8. Smith, A.L. 1991. *The Analytical Chemistry of Silicones.* New York: John Wiley & Sons.

9. Mark, H.F.; Kroschwitz, J.I. 1985. *Encyclopedia of Polymer Science and Engineering*, 2nd Ed. New York: Wiley.
10. Cuddihy, E.F. April 13, 1978. Encapsulation Materials Trends Relative to 1986 Cost Goals, JPL Document No. 5101-61, Jet Propulsion Laboratory, Pasadena, California.
11. Araki, K.; Kondo, M.; Uozumi, H.; Yamaguchi, M. 2003. "Material Study for the Solar Module Under High Concentration UV Exposure," Third World Conference on Photovoltaic Energy Conversion May 11–18, Osaka, Japan.
12. Bruins, P.F. 1970. *Silicone Technology*. Easton, PA: Interscience.
13. Mark, J.E. 1999. *Polymer Data Handbook* Oxford: Oxford University Press.
14. Pern, F.J.; Czanderna, A.W.; Emery, K.A.; Dhere, R.G. 1991. Weathering Degradation of EVA Encapsulant and the Effect of Its Yellowing on Solar Cell Efficiency. *Conference of the 22nd IEEE Photovoltaic Specialists Conference* 1: 557–561.
15. Pern, F.J. 1996. Factors that Affect the EVA Encapsulant Discoloration Rate upon Accelerated Exposure. *Solar Energy Materials and Solar Cells* 41–42: 587–615.
16. Jin, J.; Chen, S.; Zhang, J. 2010. UV Aging Behavior of Ethylene-Vinyl Acetate Copolymers (EVA) with Different Vinyl Acetate Contents. *Polymer Degradation and Stability* 95: 725–732.
17. Czanderna, A.W.; Pern, F.J. 1996. Encapsulation of PV Modules Using Ethylene Vinyl Acetate Copolymer as a Pottant: A Critical Review. *Solar Energy Materials and Solar Cells* 43: 101–181.
18. Kempe, M.D.; Jorgensen, G.J.; Terwilliger, K.M.; McMahon, T.J.; Kennedy, C.E.; Borek, T.T. 2007. Acetic Acid Production and Glass Transition Concerns with Ethylene-Vinyl Acetate Used in Photovoltaic Devices. *Solar Energy Materials and Solar Cells* 91: 315–329.
19. Gebelein, C.G.; Williams, D.J.; Deanin, R.D. 1983. *Polymers in Solar Energy Utilization*. Washington DC: American Chemical Society.
20. Holden, G.; Kricheldorf, H.R.; Quirk, R.P. 2004. *Thermoplastic Elastomers*, 3rd Ed. Munich: Hanser.
21. Massey, L.K. 2007. *The Effects of UV Light and Weather on Plastics and Elastomers*. Norwich, NY: William Andrew.
22. Jorgensen, G.J.; Terwilliger, K.M.; DelCueto, J.A.; Glick, S.H.; Kempe, M.D.; Pakow, J.W.; Pern, F.J.; McMahon, T.J. 2006. Moisture Transport, Adhesion, and Corrosion Protection of PV Module Packaging Materials. *Solar Energy Materials and Solar Cells* 90: 2739–2775.
23. Lalguna, B.; Sánchex-Friera, P.; Ropero, F.; Gil, J.F.; Alonso, J. 2008. "Comparison of Moisture Ingress in PV Modules with Different Backsheets Using Humidity Sensors." Presented at the 23rd European Photovoltaic Solar Energy Conference, September, Valencia, Spain.
24. Wohlgemuth, J.; Shea, S. April 2002. PVMaT Improvements in the BP Solar Photovoltaic Module Manufacturing Technology NREL/SR-520-32066.
25. Harpers, C.A. 1975. *Handbook of Plastics and Elastomers*. New York: McGraw-Hill.
26. Cuddihy, E.; Coulbert, C.; Gupta, A., Liang, R. 1986. Electricity from Photovoltaic Solar Cells: Flat-Plate Solar Array Project Final Report. Volume VII: Module Encapsulation JPL Publication 86-31 NASA, Springfield, VA.
27. Oreski, G.; Wallner, G.M. 2005. Delamination Behavior of Multi-Layer Films for PV Encapsulation. *Solar Energy Materials and Solar Cells* 89: 139–151.

28. Realini, A.; Burá, E.; Cereghetti, N.; Chianese, D.; Rezzonico, S.; Sample, T.; Ossenbrink, H. 2002. Study of a 20-Year-Old PV Plant (MTBF Project), Annual Report 2002-Swiss Federal Office of Energy.
29. Odian, G. 1991. *Principles of Polymerization*, 3rd Ed. New York: John Wiley & Sons, Inc.

4

Polymer Processing Techniques Used in Photovoltaic Packaging and Balance of Systems (BOS) Component Fabrication

4.1 Common Polymer Processes for Photovoltaic Packaging and BOS Components

Thermoplastics sold as pellets or sheets are not in the final forms required for module assembly. Instead, they must be formed into the required shape during a secondary processing step. Injection molding is the most common processing technique for photovoltaic (PV) module components. Secondary suppliers contracted by the PV manufacturers (Table 4.1) commonly perform injection molding.

Lamination and adhesive dispense are processing techniques performed directly by the PV manufacturers. Lamination is the process typically used to encase the PV cells in a thermoplastic encapsulant. Adhesives are used to assemble the frames around the encapsulated cells or attach junction boxes to the underside of the module. Elastomers and thermoset adhesives are pumped from drums, mixed in the appropriate ratio, and dispensed between mating interfaces during panel framing.

4.2 Polymer Viscosity

Polymer processing is the study of the conditions and techniques used to convert polymeric resins into their desired shape. Each processing technique must be optimized to minimize noncompliant parts. Optimization requires acknowledgement of the correlation between applied stress (e.g., pressure and temperature) and corresponding behavioral change (e.g., viscosity).

Viscosity is the most important material property for process optimization, because it is a measure of polymeric flow. Viscosity is a function of

TABLE 4.1

Photovoltaic (PV) Packaging Components and Common Processing Methods

PV Component	Lamination	Adhesive Dispense	Injection Molding
Encapsulation of PV cells	•		
Frames		•	•
Junction box		•	•

intrinsic properties, such as molecular weight, and extrinsic conditions, such as temperature, pressure, and shear stress.

The weight average molecular weight of the polymer influences the observed melt viscosity, measured in newton-seconds per square meter $((N \bullet s)/m^2)$, also known as pascal-seconds (Pa-s). Zero shear force viscosity (η_o) is directly proportional to the weight average molecular weight (M_w), in grams per mol (g/mol), through the material coefficient (K), measured in moles-newtons-seconds per square meter-grams $((mol \bullet N \bullet s)/(m^2 \bullet g))$ (Equations 4. 1 and 4.2).

$$\eta_o = KM_w \quad \text{for } M_w < M_c \tag{4.1}$$

and

$$\eta_o = KM_W^{3.4} \quad \text{for } M_w > M_c \tag{4.2}$$

At values below the critical molecular weight (M_c), the polymer chains are not long enough to entangle each other. In this region, there is a linear relationship between the weight average molecular weight and polymer viscosity. Above the critical molecular weight, the polymer chains entangle and inhibit flow. In this regime, the viscosity increases proportional to the weight average molecular weight raised to the 3.4 power. These characteristic equations hold for all polymers, but the critical molecular weight is dependent on the specific polymer chemistry. For most polymers, it is between 5000 and 15,000 g/mol [1].

When pressure is applied, polymer viscosity increases. Increasing pressure decreases the free volume in the polymer by pressing chains out of their equilibrium configuration and into closer proximity with each other. This decreased internal volume makes it more difficult for them to slide past each other, manifesting in an increase in viscosity. Increasing pressure has the same effect on viscosity as decreasing temperature.

As the polymer temperature is increased and thermal transitions are surpassed, the polymer chains increase in kinetic energy. With increased energy, they are able to slide past one another. Therefore, an inverse relationship exists between viscosity (η) and temperature (T). Specifically, an increase in temperature will decrease viscosity.

Viscosity can be fit to an Arrhenius equation using the empirical constants (A), measured in $(N \bullet s)/m^2$, and activation energy (E_a), measured in joules per mole (J/mol), along with the universal gas constant ($R = 8.31451$ J/(K • mol)) and temperature, in Kelvin (Equation 4.3):

$$\eta = A \, \exp(-E_a/RT) \tag{4.3}$$

The Arrhenius equation fits empirical data for most molecular weights and temperature intervals.

Under certain circumstances, polymer viscosity is dependent on the experimental shear rate ($\dot{\gamma}$), measured in inverse seconds (s^{-1}). At low shear rates ($\dot{\gamma} < 10 \, s^{-1}$), polymers typically exhibit a Newtonian plateau, meaning viscosity is independent of shear rate. At higher shear rates ($10 \, s^{-1} < \dot{\gamma} < 10^3 \, s^{-1}$), polymers will exhibit a power law dependence. This relationship, also known as the power law of Ostwald and de Waele, defines viscosity as a function of shear rate raised to the dimensionless power law index (n) and multiplied by the consistency index (m), measured in ($N • s^n$)/m^2 (Equation 4.4):

$$\eta = m\dot{\gamma}^{n-1} \tag{4.4}$$

Most thermoplastics exhibit a shear thinning behavior, meaning the power law index is less than one and viscosity decreases with an increase in shear stress. The lower the viscosity, the smaller is the stress required to induce a change in flow.

4.2.1 Viscosity Measurement

Rheology is the science of viscosity measurement, and rheometers are the instruments used to measure viscosity. These instruments are differentiated based on the stress field placed on the material to induce flow. The most common is a shear stress rheometer utilizing a cone and plate geometry. The polymer is placed between a shallow rotating cone, with a typical cone angle of one degree, and a stationary plate. The measured viscosity (η) is defined as shear stress (τ), measured in force per unit area (N/m^2), divided by the rotational shear rate (κ), measured in inverse seconds (s^{-1}) (Equation 4.5):

$$\eta = -\frac{\tau}{\kappa} = \frac{(3T/2\pi R^3)}{(\Omega/\alpha)} \tag{4.5}$$

The shear stress is derived from the torque (T), measured in newton–meters ($N • m$), and flow radius (R), measured in meters (m), and shear rate is defined by the angular speed (Ω), measured in degrees per second, divided by the angle between the cone and plate (α), measured in degrees. The experimental output is a graph of viscosity (η) as a function of shear rate.

Adhesive dispense is a packaging process commonly optimized with shear rheology. Arguably, the most critical processing parameter for automated dispense is dispense accuracy. Dispense amount is directly related to material viscosity. If the viscosity is too low, the dispense amount will be too high, erratic, and hence, wasteful. If the viscosity is too high, the dispense amount is too low or the dispense lines could clog, requiring machine maintenance. Viscosity curves generated under isothermal conditions at various shear rates allow the engineer to set the appropriate line temperature and flow rates to optimize automated equipment. The processing window is typically set in the center of the power law region where the adhesive components exhibit a shear thinning behavior.

Outside of this context, viscosity measurements have limited industrial value. Instead, it is a common industrial practice to measure the melt flow index (MFI) to predict processing behavior. The MFI is performed on solid polymers or viscous oils at the manufacturer's suggested processing temperature. MFI is measured by placing the polymer in a controlled thermal reservoir under a weight. The weight used in the test is included in the specified test conditions, ASTM D1238 [2], but a 2- to 3-kilogram weight is typical. The amount of material, in grams, that flows through a capillary to the bottom reservoir within 10 minutes defines the melt flow index (g/10 min).

MFI, not viscosity, is typically reported on technical data sheets. There is an inverse relationship between MFI and viscosity, with a high MFI indicating a low viscosity and a low MFI indicating a high viscosity. Polymers used for adhesives and lamination have the highest MFI and can easily flow with limited pressure. Injection molding is typically performed on thermoplastics with a low MFI, requiring high packing pressure and high temperatures to mold the polymer into the desired shape.

4.3 Lamination

Laminators can be purchased off the shelf from a number of manufacturers. Commercial laminators are divided into an upper and lower chamber, and both are connected to a vacuum pump and separated from each other by a flexible membrane. The lower chamber houses a platen that can be heated and cooled to the desired processing temperatures (Figure 4.1).

Lamination processing steps include module lay-up, evacuation, polymerization, cool down, repressurization and ejection. The process flow starts with module layup. The encapsulant and backsheet are cut from rolls. The backsheet is placed on the bottom platen, and the encapsulant sandwiches the PV cells and electrical contacts. After the glass is placed on top of the stack, the upper chamber is closed and evacuated while the temperature is elevated. Evacuation is a requirement for air removal from all material interfaces. The pressure from the top chamber is increased as the temperature is raised above the encapsulant's melt temperature. After evacuation, polymerization begins while the polymer is held under pressure and heat.

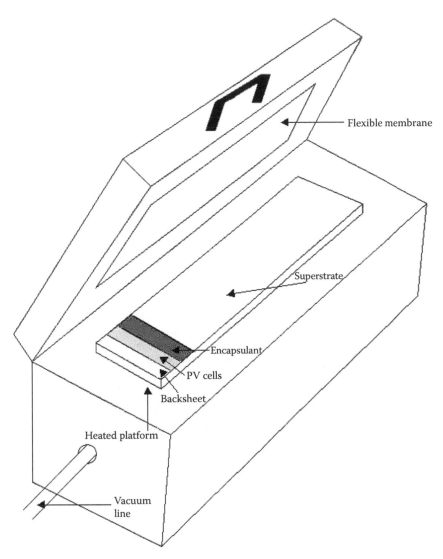

Flexible membrane

Superstrate

Encapsulant

PV cells

Backsheet

Heated platform

Vacuum line

FIGURE 4.1
A research laminator used for photovoltaic (PV) lamination.

The encapsulant will flow around the PV cells while chemically bonding to the glass and backsheet interfaces. Polymerization is the longest processing step, with a typical dwell on the order of minutes. The required polymerization temperature is specified by the polymer manufacturer's data sheet. The chamber is cooled to the equilibrium temperature and held for a few minutes before the air pressure is increased to allow for unloading. Once removed, the edge seal is trimmed to eliminate excess flashed encapsulant, and the module is moved from the encapsulation line to the framing line.

TABLE 4.2

Failure Modes, Corresponding Potential Causes and Corrections for Laminated Photovoltaic (PV) Cells

Failure Mode	Causes and Corrections
Bubbles	• Insufficient pumpdown • Air is trapped, a burp path is required (e.g., craneglass) • Temperature ramp is too fast and causing moisture or outgassing from other materials • Cure temperature is too high • Premature pressurization of top chamber • Latent pressurization of top chamber
Insufficient mechanical strength	• Encapsulant has expired or was inappropriately stored • Temperature ramp is too slow • Cure time is too short • Cure temperature is too low
Encapsulant flashing to backsheet	• Add a sacrificial layer to laminator surface • Decrease pressure
Cell breakage	• Resin temperature too high • Backfill rate too high
Yellowing encapsulant during processing	• Temperature too high • Uneven temperature gradient
Delamination	• Too much primer • Oil or residue on the glass • Water-based primer was used or assembly components were not sufficiently dry • Insufficient encapsulant cure • Primer has expired

The operator can use pressure, temperature, or time to optimize the package quality.

Vacuum pressure must be optimized to remove air and volatiles both during evacuation and during polymerization. If the processing pressure is too low or uneven, the module exhibits bubbles and delamination (Table 4.2). If it is too high, cells will break and excessive encapsulant flashing may occur.

The ramp and polymerization temperature must be optimized for the material system. These are critical processing parameters to avoid module defects and are the most common source of processing issues. If the ramp is too fast, the air and evolved gas do not have an opportunity to evacuate, causing bubble formation. Similarly, if the temperature is too high, moisture and additives in the formulation rapidly volatilize, also creating bubbles. Excessive polymerization temperatures can cause thermal degradation, and insufficient temperature or insufficient dwell time limits crosslinking. In both extremes, the mechanical properties of the encapsulant are compromised.

4.4 Injection Molding

Injection molding converts polymer pellets into a user-defined shape. Pellets are loaded into a hopper and fed into the barrel for processing. The barrel contains a reciprocating screw divided into three zones: solids conveying, melting, and melt pumping. Each zone describes a different method of material conversion. Solids conveying represents the first few elements of the screw used to transport the material away from the hopper (Figure 4.2). Within the melting zone, the rotating screw shears the polymer against the barrel and mixes it into a viscous melt. While rotating, the screw also reciprocates laterally to build up the necessary pressure to inject polymer into the mold. The molten polymer is injected into the sprue past the runner and into the gate of a clamped mold. The mold shape is machined to the consumer's specification and is a negative impression of the final part.

There is a temperature difference between the barrel and the mold. The barrel temperature is determined by the polymer's melt temperature, if the polymer is semicrystalline, and the glass transition temperature, if it is amorphous. The mold temperature is typically held slightly above ambient and below the processing temperature in the barrel. The gradient in temperature allows for the polymer to evenly flow into the mold, and it slowly begins to solidify prior to ejection.

There are a number of failure modes that can yield noncompliant parts. Although it is unlikely for PV manufacturers to directly perform injection molding, understanding these failures can facilitate their discussions with suppliers. These failures can be categorized based on the area in the tool where the error occurs. The hopper, gates, or molds are the most common processing areas for failure.

If the hopper and mold are not completely cleared between runs, the part can become contaminated. A small percentage of contaminate resin can change the polymer's haze, coloration, and mechanical properties. The contaminate resin is usually immiscible, causing it to sequester into isolated areas of the part. Areas with slightly different refractive indices will cause

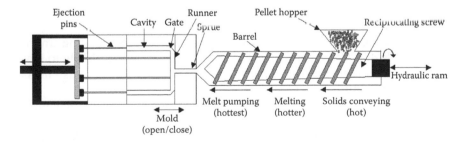

FIGURE 4.2
The injection molding machine.

TABLE 4.3

Failure Modes, Corresponding Potential Causes and Corrections for Injection-Molded Parts

Failure Mode	Causes and Corrections
Flash	• Injection process window is not optimized (speed, pressure, or time) • Inappropriate clamp pressure on the mold
Part is prematurely cracking	• Check for contamination • Check the polymer additives are not degraded
Race tracking	• Polymer flow is not optimized out of the gate
Gate blushing	• Gate angle might not be aligned • Processing conditions are not optimized
Voids	• Material viscosity is too low due to incorrect processing temperature • Packing pressure is not optimized during molding
Discoloration or black spots	• Decomposition of the polymer because the screw speed is too high, temperature is too high, or back pressure is too high • Improper flow in the channels • Contamination from another polymer • Excessive fines
Bubbles	• Too much moisture in the polymers • Polymer decomposition • Inappropriate venting • Screw is decompressing inappropriately, entrapping air • Excessive fines
Surface finish issues	• The gate is blocked • There is a contaminate on the surface of the molds
Splay	• Polymer is not appropriately dried • Inappropriate back pressure
Weld lines	• Injection velocity is not optimized • Mold is not filling uniformly • May not be venting properly
Warp	• Wall thicknesses are not appropriately designed • Cool time before ejection is too low • Gate is partially clogged

haze. In addition, these areas can act as stress concentrators inducing embrittlement (Table 4.3). If the contaminate is another resin, it is unlikely to have the same melt temperature as the part. Contaminate polymers can usually be identified by raising the part temperature to verify it melts evenly.

The polymer could exhibit embrittlement, haze, or splay as a result of moisture entrapment. If the polymer is not properly dried prior to molding, the part will bubble when the retained moisture vaporizes in the mold, leading to the aforementioned compromises in integrity. The manufacturer's data sheet will make recommendations for drying the resin prior to processing.

TABLE 4.4

Relevant Properties for General Purpose, Reinforced, and Flame Retardant DuPont Rynite®

Processing Properties	General Purpose	Reinforced	Flame Retardant
Shrinkage in MD (%)	0.23–0.13	0.24–0.13	0.34–0.13
Shrinkage in TD (%)	0.82–0.66	0.67–0.59	0.71–0.40
Water absorption (wt%)	0.05–0.04	0.25–0.06	0.07–0.04

Source: Data from DuPont, DuPont Rynite® PET Technical Data Sheet, 2010, http://plastics.dupont.com/.

These recommendations specify the drying temperature and duration; however, special care should be taken to verify dryness for grades prone to absorption. Typically, reinforced thermoplastics will exhibit higher moisture absorption. For instance, some reinforced Rynite® grades have a ~6 times higher water absorption than other grades (Table 4.4) [3].

The flow into the gate can create a number of defects. The gate is the small orifice the melt passes through to enter into the mold. This change in diameter creates an aligned area of flow resulting in orientation of the polymeric chains. Unstable flow from the gate can create concentric circles on the part called race tracking. Similarly, a change in surface finish and gate blushing represents unoptimized flow, creating localized polymer chain alignments. These aesthetic anomalies translate into weak areas of the part susceptible to premature mechanical failure. Gate alignment, temperature, and pressure can be optimized to eliminate these defects.

When the mold does not fill properly, the final part can exhibit weld lines, short shot, or both. Weld lines form when two flow fields meet and incompletely merge before cooling. They are visually displeasing and decrease mechanical integrity. These lines can act as stress concentrators during mechanical deformation and thereby weaken the molded part. Various polymers will have different responses to weld lines; however, decreased yield and fatigue strength are the most common. To avoid weld lines, mold temperatures can be increased to decrease viscosity and thereby increase mixture at flow fronts. A more serious issue is short shot. A short shot describes missing features due to incomplete mold filling. To improve mold flow, the viscosity is decreased with an increase in temperature or packing pressure. However, if the temperature is too high or if the packing pressure is increased too much, then the material will flash out of the clamped mold, creating an additional part feature. In addition, high temperatures and pressures can increase preventive maintenance requirements due to mechanical wear.

An alternative to increased temperature and increased packing pressure is to use viscosity modifiers. These additives decrease viscosity, resulting in a more uniform flow from the barrel into the mold. This allows for a more complete filling and mixing of flow fronts while decreasing wear on the processing equipment.

The polymer's flow dictates the location with greatest orientation in the molded part. As the molten polymer fills the mold, the largest quantity of polymer chains flows laterally with the machine direction. A smaller quantity will spread out perpendicularly from the gate. This latter flow direction is described as transverse to the machine. Failure to rectify this orientation prior to ejecting the part from the mold can result in a molded-in stress.

Both amorphous and semicrystalline polymers can exhibit molded-in stress. For amorphous polymers, the stress is confined to the surface with no orientation in the center of the part. In contrast, for semicrystalline parts, there is varying orientation throughout the part thickness. A semicrystalline polymer will exhibit a highly oriented skin layer aligned in the direction of flow, a transition region called the transcrystalline zone with grains forming along the temperature gradient and a spherulitic zone in the center of the part with random crystal orientation.

Molded-in stress should be avoided because it compromises mechanical properties and induces warp. The orientation at the surface is under compression, and the bulk is under tension. Therefore, microscopic orientation manifests as a decrease in macroscopic impact strength. The shrinkage and resultant warpage are always higher in the machine direction. Polymers with reinforcement particles typically exhibit lower percent shrinkage. For example, reinforcement grades of Rynite exhibit shrinkage in a tighter distribution, as compared to unreinforced grades (Table 4.4) [3]. The amount of shrinkage can also be managed in the process by ensuring even wall thickness and even temperature profiles. Fortunately, the stress is a metastable state, and it is reversible with an annealing treatment.

Thermocouples are placed inside the barrel cavity and the mold to verify desired temperatures are met for each step of the process. Faulty thermocouple placement or incorrect readings can result in a decrease in part yield. Common issues include discoloration, streaking, or void formation due to polymer thermal degradation under excessive temperatures. Polymer degradation during processing can be verified with an increase in the part MFI relative to the unprocessed pellets. In addition, abnormally high temperatures can cause blooming or gloss variation as additives migrate to the part surface.

The cycle time for the process is constrained by the time required for the part to cool before it is ejected from the mold. When the part is ejected prior to solidifying, it will warp. Additionally, the pushpins used to eject the part from the mold can leave a permanent impression.

The part can also warp due to adhesion to the mold surface. To counter this effect, formulations include mold release agents. These additives are part of the commercial formulation but are immiscible with the polymer at melt temperatures. The mold release agents sequester to the mold–polymer interface, lowering the injection pressure required to remove the part. This limits part deformation provided the material has solidified prior to mold ejection.

4.5 Adhesive Dispense

An adhesive dispenser is used to meter thermosets or elastomers onto mating surfaces. For an automated process, the one-part chemistry is loaded in a drum. Hydraulic presses or gas pressure reduces the internal drum volume, pushing the viscous material into the hoses, through the flow controller, and into the dispense head (Figure 4.3). In a two-part dispense, one drum contains the base and the other the catalyst. The material is dispensed in separate lines. The two components are combined in a nozzle at the tip of the dispense head (Figure 4.4).

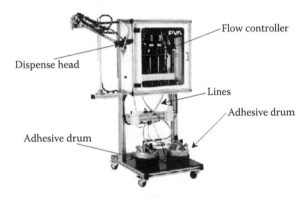

FIGURE 4.3
Automated dispense setup for adhesives. (From MX3000/MX4000 Gear Pump Meter-Mix Dispensing Systems, http://www.pva.net. With permission.)

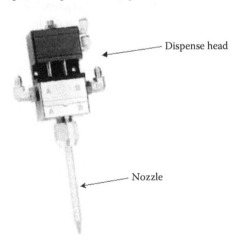

FIGURE 4.4
Dispense head for automated adhesive dispense. (From MX3000/MX4000 Gear Pump Meter-Mix Dispensing Systems, http://www.pva.net. With permission.)

TABLE 4.5

Adhesive Metering Processes and Corresponding
Descriptions

Metering Process	Description
Streaming	Nozzle is raised above the surface and the adhesive is continually dispensed in a zig-zag pattern
Extrusion	Nozzle is brought into contact with the surface and the adhesive is dispensed in a line
Metered ejection	Nozzle comes down to the surface and gently lifts off while dispensing adhesive; leaves a metered amount of adhesive behind
Spraying	Material is atomized above the surface and falls below creating a conformal coating
String dispersion	Continuous adhesive dispensed from the nozzle elevated above the substrate which creates a linear pattern below

Nozzle tips are either dynamic or static. A static mix nozzle has a combination of elements molded in place in a plastic tip. As the material is dispensed, it shears and blends as it moves around the elements. In contrast, a dynamic mix tip has rotating mix elements that turn at a programmed speed, measured in rotations per minute (rpm).

The dispense technique is based on the PV manufacturer's requirements. There are five possible processing conditions: streaming, extrusion, metered ejection, spraying, and string dispersion (Table 4.5). Each technique has a characteristic dispense pattern that defines the bond area. There is an inverse correlation between the bond area and the precision of the dispense technique.

Streaming, spraying, and string dispersion create the largest bonding area and therefore some of the highest peel strengths between two interfaces (Figure 4.5). The dispense head is elevated above the substrate, minimizing dispense precision but optimizing coverage. Streaming results in a constant zig-zag flow and bond area. Spraying and string dispersion increase bond area by evenly spreading the adhesive across the substrate's surface area.

Metered ejection and extrusion have similar mechanical requirements (Figure 4.6). Both metering and extrusion require the dispense head to come into contact with the surface, thus increasing the dispense precision. Metered ejection creates a droplet with a small surface area of adhesive defined by the droplet radius. Strings, between adhesive droplets, are a common processing instability that create an unwanted bond line (Table 4.6). Incomplete adhesive retraction before moving to the next dispense area can be improved with higher retraction forces in the dispense nozzle. Extrusion creates a thin bond line between two surfaces. It is similar to streaming except with more dispense control.

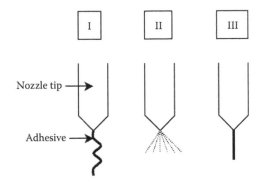

FIGURE 4.5
(I) Streaming, (II) spraying, and (III) string dispersion of adhesives out of dispense nozzles.

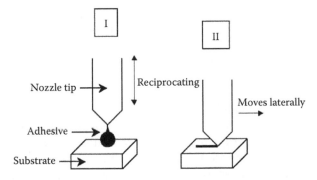

FIGURE 4.6
(I) Metered ejection and (II) extrusion of adhesives out of dispense nozzles.

Metered ejection and extrusion are the two most common adhesive dispense methods used in PV applications. Metered ejection is a common method for potting and casting, and extrusion is commonly used in module framing.

Process optimization is time and mix ratio dependent. Polymer manufacturers specify a working time on their data sheets. At the end of the working time, any stress applied to the adhesive will result in permanent deformation. The two surfaces must be joined prior to the end of the working time to ensure optimal peel strength. The cure time is the dwell required for permanent mechanical properties to manifest. One-part adhesives have longer work and cure times than two-part adhesives. However, both chemistries will exhibit abnormal cure times if component ratios are skewed or expired material is used. If complete cure never occurs, the likely cause is an incorrect mix ratio. A symptom of this issue can be abnormal pressures obstructing the flow of components. The principal causes of increased pressure are settled fillers in the dispense lines or cured adhesive clogging the dispense tip.

Potting describes a processing technique in which the adhesive is dispensed into a permanent enclosure, such as a junction box. The two most common

TABLE 4.6

Failure Modes, Corresponding Potential Causes, and Corrections for Adhesive Dispense

Failure Mode	Causes and Correction
Bubbles	• Air in dispense lines • Drum is not properly sealed • Material has reached its shelf life and started to decompose
Strings or dripping between metered areas of dispense	• There is no negative pressure after dispense, also termed snuff-back
There are soft spots on the adhesive after the manufacturers' specified cure time	• Ensure the proper mix tip is used (e.g., length, number of elements) • Leaky solvent from the formulation may be contaminating the lines and slowing cure • Shelf life may have expired
Abnormal cure times	• Shelf life may have expired • Improper mix due to incorrect tip choice • Inappropriate metering
Abnormal pressures required for dispense	• Filler particles have separated from the polymer

processing failures are bubbles and shrinkage. There is no optical requirement to remove bubbles from junction boxes, but the processing chamber is usually evacuated during cure to level the pottant's surface and to force removal of deleterious by-products emitted during cure. Cure shrinkage must be minimized to prevent unwanted stress on the electrical components, and it is commonly controlled by choosing a low shrinkage chemistry.

Casting is a typical processing technique used for electronic applications. However, this technique has not had widespread commercial PV success. Even though it is rarely used today, most of the PV panels produced in the 1960s for the National Aeronautics and Space Association (NASA) satellites were manufactured by silicone casting. A layer of silicone was applied to the glass, the PV cells were placed into the silicone, and then a second layer of colored silicone was added as a backdrop. The entire device would be placed in a vacuum chamber to evacuate air bubbles. A box was used to provide physical shape to the silicone during cure, and it was removed after cure was complete. It was customary to use a two-part silicone to precisely control cure time. This method of manufacturing was not utilized in module commercialization because it was not a continuous production process, and silicone's material costs were too high.

Recognizing these limitations, the silicone industry has made some innovations. Dow Corning (Midland, Michigan) recently unveiled a pilot operation of a continuous process. The cure rate has been optimized using their PV-6100 series formulations and throughput has increased to less than two modules per minute. Additional details of the manufacturing process have not been publicly disclosed.

References

1. Roy, S.K.; Chanda, M. 2007. *Plastics Technology Handbook,* 4th Ed. Boca Raton, FL: CRC Press.
2. ASTM Standard D1238-10, 2004. 2000. "Standard Test Method for Melt Flow Rates of Thermoplastics by Extrusion Plastometer." ASTM International, West Conshohocken, PA. DOI: 10.1520/D1238-10, www.astm.org
3. DuPont. 2010. DuPont Rynite® PET Technical Data Sheet. http://plastics.dupont.com/

References

5

Economic Theory and
Photovoltaic Packaging

5.1 The First U.S. Energy Crisis

The photoelectric effect is the physical phenomenon of turning light into electricity. It was discovered by Alexander Edmond Becquerel almost two centuries ago in 1839. However, it was not until 1954, over a century later, that the first silicon photovoltaic (PV) cell was fabricated by Bell Labs [1]. Despite this proof of concept, PV power remained cost prohibitive to the general public with an asking price of ~$1000/watt in 1955. A number of American companies noted the potential for commercialization with increased cell efficiency and lower costs; however, even without these improvements, PV energy quickly found a niche market for government applications.

During this same decade, in 1958, the National Aeronautics and Space Administration (NASA) was formed as a result of the National Aeronautics and Space Act signed into law by President Dwight D. Eisenhower. Due to the successful launch of the Russian satellite, Sputnik I, NASA was chartered to perform and manage space exploration and research designed to offset foreign threats to national security. Lightweight cells were an attractive energy source for both satellites and space shuttles that were vital to accomplishing the agency's mission. Launched in 1958, Vanguard I was the first U.S. satellite to utilize the silicon cells developed a few years earlier [2].

Between the early 1950s and late 1970s, photovoltaic modules were used for specialized government applications that could absorb their high manufacturing and material costs. In 1976, crystalline silicon modules commanded a retail price of $51/watt [3], still beyond the reach of the average American household. Thus, PV energy had failed to be widely commercialized.

A catalytic event was required to change the energy market and make PV a competitive resource. That catalysis occurred in 1973 when the United States underwent its first energy crisis resulting from an oil embargo from the Middle East. Between 1973 and 1974, oil prices quadrupled (Figure 5.1), and the United States scrambled to decrease dependency on foreign oil. In an act of leadership, President Jimmy Carter became the first U.S. president to

FIGURE 5.1
Oil prices in the United States from 1970 to 2009. (Data from EIA [Energy Information Administration], U.S. Department of Energy, Washington, DC, 2008.)

place PV modules on the White House. Recognizing American households could not afford the $51/watt for PV energy, he encouraged them to exercise energy conservation and turn down their thermostats.

In hindsight, economic surveys conducted in the late-1970s indicated the United States was vulnerable to an oil embargo [4]. This vulnerability stemmed from U.S. goods and services engineered for decades around high oil and gasoline consumption. In economic terms, this created an inelastic demand. Simply put, the U.S. consumer had purchased fuel-inefficient cars and homes that could not be easily modified when oil prices rose. Long, rationing lines at gas stations have become an iconic moment that defines the social temperature of that decade. Despite this frustration, consumer surveys indicated a 10% increase in oil prices only resulted in a 2% to 4% decrease in demand. Homes and automobiles were too expensive to easily be replaced with energy-efficient models. There was no short-term market for renewable energy [4]. Even with quadrupling oil prices, PV was still too expensive. The challenge was how to make PV energy competitive in the short run.

In response, the American Chemical Society (ACS) called together top researchers and public officials to Cherry Hill, New Jersey, to discuss the

U.S. dependency on foreign oil [5]. As part of this famous conference, now known as the Cherry Hill Conference, the Department of Energy (DOE) chartered the Jet Propulsion Laboratory (JPL) with investigating PV energy as a commercial option. As the principal consumer for PV technology, NASA was prepared for researching the industrial requirements for the rapid transfer of PV from the private to the public sector. To execute this directive, JPL procured commercial PV modules for internal testing and evaluation. Performance and reliability characteristics of the current "state of the art" were studied to gain insight into the requirements to decrease costs. These studies were referred to by the JPL as *Block Buys*. JPL continued to issue annual reports of their finding into the mid-1980s. However, by the end of his term, Carter had decided the future development and technology transfer of renewable energy should be the focused effort of a new branch of the Department of Energy, named the Solar Energy Research Institute. This branch of the DOE would, in 1991, be renamed the National Renewable Energy Laboratory (NREL).

Oil prices nearly quadrupled again during the Iran–Iraq war, and public outcry reached a crescendo by the close of the 1970s decade. To emphasize its commitment for finding a solution, the federal government created legislation to make alternative energy more competitive. Congress passed the National Energy Act of 1978 providing a framework for future investment in renewable energy. These bundled policies included government funding and tax incentives aimed at increasing the U.S. renewable energy consumption to 20% of the national total by the year 2000. Most notably, the policy included the Public Utility Regulatory Policies Act (PURPA) and the Energy Tax Act. PURPA created a demand for renewable energy by forcing utilities to purchase a portion of their energy from renewable sources. The government subsidized these prices at a special rate so renewable energy was competitive with oil and coal. However, enforcement of PURPA was left to the individual states, making the legislation uneven and ineffective [6]. The Energy Tax Act included income tax credits for the purchase of fuel-efficient automobiles and alternative energy for residential and commercial buildings. This latter legislation was designed to increase the elasticity of energy demand. Using tax credits, U.S. consumers could now afford energy-efficient cars and homes.

After losing the election in 1980, the Carter administration's renewable energy initiatives were quickly reversed. During his term, Ronald Reagan removed the White House's PV modules. This was a symbolic gesture that set the tone for the new administration's attitude toward PV technology. The 1980 to 1988 terms were full of policy reversals and reduced spending for renewable energy. The incentives passed under the National Energy Act timed out without renewal. More overt policy reversals included reducing the Department of Energy's budget for energy conservation and renewable energy programs. The popular view of this decade was to allow market forces, rather than government intervention, to determine the viability of renewable energy [7]. Although PV programs evaporated and PV companies fell into bankruptcy, nuclear energy and coal thrived during this period as a means to offset foreign oil dependency [6].

5.2 The Current Energy Crisis

After the U.S. invasion of Iraq in 2003, oil prices began to steadily increase due to the rising Middle East turmoil. By 2005, the U.S. price for oil had doubled the average cost in the 1990s (Figure 5.1). In response, the U.S. Congress passed the Energy Policy Act of 2005. Similar to past decades, this new legislation was meant to increase energy diversification and improve energy efficiencies through a series of tax incentives and loan guarantees for alternative energy. However, the policy was widely criticized for emphasizing U.S. exploration of new oil reserves and neglecting renewable sources.

Regardless, from 2005 to 2008, there was a linear increase in the number of U.S. PV companies. By 2008, there were 66 PV companies, more than double the 29 companies in existence in 2005 (Figure 5.2). Both government initiatives and private-sector investments bolstered this growth. As the biotechnology and information technology investments started to wane, renewable energy became the new darling of venture capital firms [9]. The investment in renewable energy had doubled by the first economic quarter of 2010, tying for the second highest amount of venture capital investment (Figure 5.3).

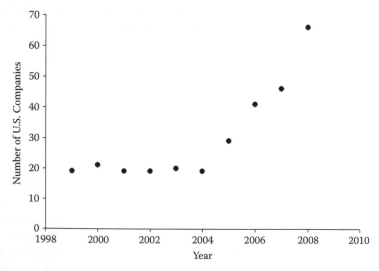

FIGURE 5.2
Number of U.S. solar companies between 1999 and 2008. (Data from EIA [Energy Information Administration], Table 3.1: Annual Shipments of Photovoltaic Cells and Modules, 1999–2008. U.S. Department of Energy, Washington, DC, 2010.)

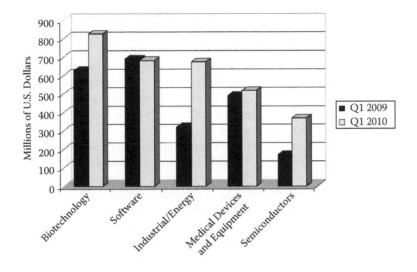

FIGURE 5.3
Venture capital investments in renewable energy start-ups. (Data from National Venture Capital Association [NVCA], Investment Activities—Top Industries Q1 2010, 2010, http://www.nvca.org/.)

In late 2007, the oil prices quadrupled the 1990 average, spurning public outcry for energy reform. This became a central part of Barack Obama's platform for election to the U.S. presidency. As the credit markets tightened and gasoline prices rose, public interest in renewable resources once again swelled. After winning the election, President Obama signed the American Recovery and Reinvestment Act in February 2009 to rejuvenate the lagging U.S. economy. The bill was meant to increase job growth and targeted the renewable energy market to create this opportunity. This new federal bill is projected to create a 37% increase in current renewable energy by 2015, a modest 9% to 10% of the total U.S. energy consumption. By 2025 there is an expectation that renewable energy will be 25% of the total U.S. consumption. To create this growth, the stimulus provides $45.1 billion for renewable energy incentives, grants, and loan guarantees over the next 10 years [8]. It remains to be seen if the technological development spurred by these investments will be enough for PV energy to become competitive in the United States.

5.3 Technology Development Theory and Photovoltaic Energy

For more than three decades, NRL has monitored PV cell efficiencies for the various semiconductor technologies. The thin-film technologies (e.g., copper indium gallium diselenide [CIGS], amorphous silicon [a-Si], and cadmium telluride [CdTe]) generally have the lowest efficiencies. Crystalline silicon PV cells have efficiencies slightly higher than thin-film technologies but use between 20 and 100 times more semiconductor material. These two classifications dominated the industry in the mid-1980s. At that time, theoretical physicists predicted PV cells would not exceed 22% efficiency. However, with the discovery of multijunction PV cells, efficiencies have now exceeded 40%. Due to the emergence of this new technology, theorist predictions have increased to a maximum efficiency of 58% to 70% [10].

Regardless of chemistry, the entire industry has seen more than an order of magnitude improvement in PV cell efficiency between 1976 (~2%) and 2010 (~41.6%) [11]. When each PV cell chemistry is viewed independently, the improvements are not as startling. Taking amorphous silicon PV cells as an example, the efficiency improved from ~2% reported by RCA in 1976 to ~12% reported by United Solar in 2009 (Figure 5.4). Economists use technology development theory to explain the disparity between the growth of the entire industry and these individual technological innovations.

Economists describe technology development as a sigmoid curve, sometimes referred to as an S-curve. The S-curve is formed from graphing technological advancements as a function of time. Economists have assigned significance to each portion of the curve. First the technology goes through a new invention stage. During this time, there is slow growth with extensive funding and little performance gain (Figure 5.5). This is followed by an exponential growth phase known as technology improvement. During this phase, fewer resources are required because scientists now understand the underlying physical phenomena required for advancement. Finally, the growth levels off when the technology reaches a physical limit preventing further development. The technology has reached maturity and enters an aging phase [12].

These advancements in PV cell efficiency can be described by sigmoidal expressions. Gompertz and logistic curves are the most popular in economic theory. First published in 1825 and used by economists since 1903, the equations describe rising exponential growth with log linearity at the beginning and end of the curve. Taking the Gompertz expression as an example, the highest and lowest plateaus are described as $L + x$ and x, respectively (Equation 5.1):

$$y = x + Le^{-e^{-b(t-t_c)}} \tag{5.1}$$

The rate of rise is dictated by the variable b found with the goodness of fit to the technology performance data (y) graphed as a function of time (t) [13,14]. The inflection point for growth occurs at a specific time interval (t_c) in the technology improvement phase. When applied to advancements in PV cell efficiency over the past four decades, the saturation limit is projected to be 48.5% with an inflection point in 1984 (Table 5.1, Figure 5.6). The poor fit

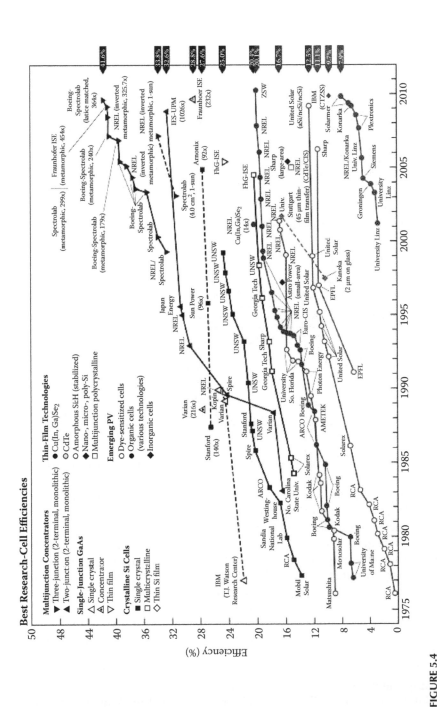

FIGURE 5.4

National Renewable Energy Laboratory (NREL) plot of the cell efficiencies of various semiconductor technologies from 1975 to 2010. (From NREL [National Renewable Energy Laboratory], National Center for Photovoltaics, Golden, Colorado, 2010, http://www.nrel.gov/. With permission.)

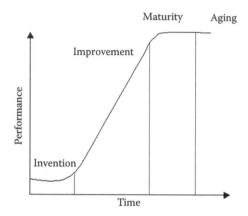

FIGURE 5.5
Explanation of the sigmoid technology curve.

TABLE 5.1

Fitting Coefficients for a Gompertz Curve of
Photovoltaic (PV) Cell Efficiencies from 1975 to 2010

Fitting Variable	Efficiencies 1975 to 2002
x	−8.0
L	56.5
b	0.08
t_c	1984
R^2	0.952
χ^2	10.47

suggests PV cell technology has not completely matured, meaning more
technological advancements could theoretically occur.

Often a single sigmoidal curve does not accurately describe industrial performance. Instead, a technology curve is composed of a number of smaller
sigmoidal curves with each curve describing a new innovation [12]. There
are two possible growth paths defined by the time between innovations. One
scenario is that a number of ideas are created during the invention stage and
simultaneously pursued. This leads to a series of overlapping S-curves for
each innovation (Figure 5.7[I]). A second scenario occurs when there are fewer
resources devoted to technology development. In this instance, one idea is
pursued until it reaches maturity. Consumer demand starts to decrease, and
the industry responds with a slightly new innovation described by a new
growth curve (Figure 5.7[II]).

When PV cell efficiency is used as the technology performance metric,
the advancements over the recent decades can be visualized as a series of
sigmoidal curves. The NREL plot can be reinterpreted to identify the three

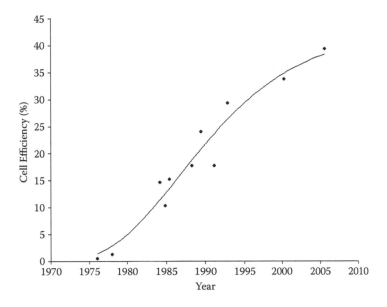

FIGURE 5.6
Gompertz fit to solar cell efficiencies from 1975 to 2010.

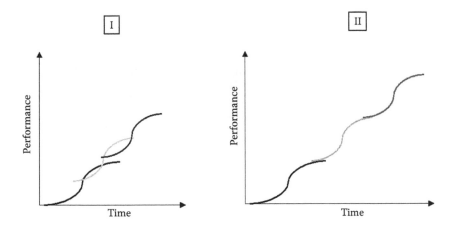

FIGURE 5.7
(I) A number of innovations are simultaneously pursued, leading to simultaneous growth.
(II) Ideas are pursued one at a time, leading to a series of end-to-end growth curves.

changes in material innovations that have led to today's highest single cell efficiencies of 41.6% (Figure 5.8).

Thin-film, crystalline, and multijunction cells are three overlapping technologies describing the industrial innovation in PV (Figure 5.8). However, based on the statistical values of R-squared and chi-squared, single crystalline silicon cells and thin-film technologies are the only good fits to the

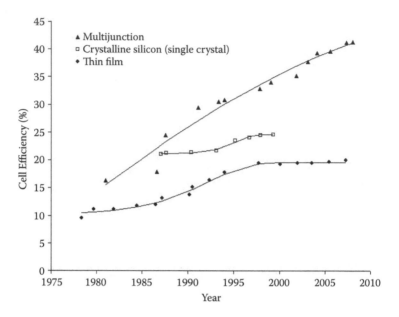

FIGURE 5.8
Fitted Gompertz curves to thin-film, crystalline silicon, and multiple junction solar cell technology between 1975 and 2010.

TABLE 5.2

Fitting Coefficients for the Gompertz Curves for Individual PV
Cell Technologies from 1975 to 2010

Fitting Variables	Thin Film	Crystalline Silicon	Multijunction
x	19.6	24.6	48.5
L	−9.5	−3.4	−122.1
t_c	1992	1995	1971
b	−0.24	−0.62	−0.02
R^2	0.99	0.99	0.96
χ^2	0.22	0.05	3.23

Gompertz prediction (Table 5.2). Upper limits of 10.1% efficiency for thin-film
and 21.2% efficiency for crystalline silicon appear to have been reached, and
the majority of technological advancements for both occurred prior to the
mid-1990s. The poor fit of multijunction technologies suggests these innovations have not reached their full maturity. This possibility is mirrored in
the abundant academic literature focused on continued efficiency improvements for this technology.

Note these efficiencies are of a single PV cell and not the packaged
module. NREL's plot contains individual cell efficiencies demonstrated on

a few test substrates synthesized in controlled academic or industrial laboratories. For instance, thin film single cell efficiencies (19.5%) are approximately two times higher than a packaged module (10%) containing the same CIGS chemistry [15]. This is not necessarily a comment on packaging integrity, but it is a demonstration of the current limitations in scalability of PV cell synthesis. The single cell efficiencies are ideal substrates, termed *champion cells* by the industry. PV manufacturers currently strive to develop processing equipment to consistently duplicate these efficiencies in high volumes.

The semiconductor chemistry has evolved, and so, too, have encapsulant formulations and processes. The principal requirement of the encapsulant material is to provide an environmental barrier and enable reliable performance of the PV cells. Technological advancements in PV cell stability and packaging integrity contribute to the warranty offered by PV manufacturers.

In 1979, when the first terrestrial PV modules were sold, they were offered with a 5-year warranty. Warranty periods have increased, and by 2010 the lower limit of the industrial norm is 25 years (Figure 5.9). A Gompertz curve fits this historical data (Table 5.3). Based on this theory, the technology has plateaued in the recent years. Therefore, a new innovation must occur for PV packaging to undergo future technological improvement. Unlike PV cell development, there is little literature on the pursuit of packaging material improvements for PV manufacturing.

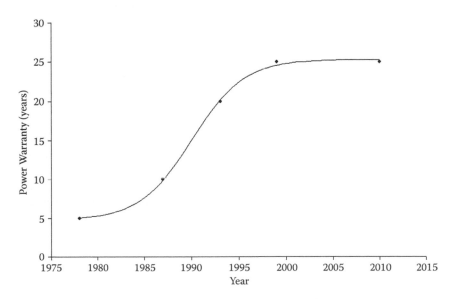

FIGURE 5.9
Fitted Gompertz curve to power warranty for solar modules between 1979 and 2010.

TABLE 5.3

Fitting Coefficients for a Gompertz Curve
to Photovoltaic (PV) Packaging Warranties

Fitting Variable	Module Warranty Data
x	5
L	20
b	0.27
t_c	1988
R^2	0.997
χ^2	1.79

5.4 Operational Optimization for Photovoltaic Companies

High costs have been blamed for the slow adoption of PV technology as an alternative energy source. These costs are directly related to a PV company's operating conditions dictated by the market. The energy market is not a perfectly competitive market. However, examining economic principles describing a perfect competition can provide important insight into the current U.S. energy market.

Following basic microeconomic principles, all companies in a perfectly competitive market, want to operate where the marginal revenue (MR) is equivalent to the marginal cost (MC) (Equation 5.2, Figure 5.10):

$$MR = MC \tag{5.2}$$

The $MR = MC$ rule states that if profit can be made by producing an additional unit, then the company should continue to operate provided there is pure competition in the market. The term *marginal* refers to the consequence of producing an additional unit of product. When marginal revenue is greater than marginal cost, the company is producing profit. Manufacturing should stop when additional units are not profitable, when marginal cost exceeds marginal revenue.

Marginal revenue is the revenue (R) gained from producing one additional unit (Q). This is the mathematical equivalent of the derivative of the revenue versus quantity curve (Equation 5.3):

$$MR = \frac{dR}{dQ} \tag{5.3}$$

Revenue (R) is equal to the price (P) of the product times the number of units sold (Q) (Equation 5.4):

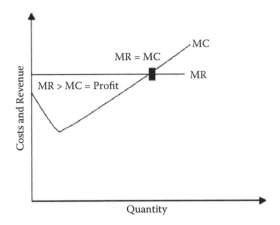

FIGURE 5.10
Typical marginal cost and revenue curves for a company.

$$R = PQ \qquad (5.4)$$

Therefore, marginal revenue is just the price of the product.

The initial, fixed price of PV installations is the summation of the cost of the array, the balance of system (BOS), the batteries, the inverter, and the installation. These costs are discounted by government subsidies in the form of rebates and tax incentives (Equation 5.5):

$$Price = Array + BOS + Batteries + Inverter$$
$$+ Installation - Rebates - Tax\ Incentives \qquad (5.5)$$

The quantity, and therefore price, of the BOS, batteries, inverters, installations, rebates, and tax incentives is dependent on the size of the array purchased. This decision is based on the buyer's power consumption in terms of kilowatt-hours (kWhr). The daily power consumption is the total kilowatt-hours/day charged to the customer, defined here as the power required ($P_{required}$).

The module will not have sunlight continuously shining on it because of seasonal variations, diurnal cycling, and environmental shading. The hours of usable light are dependent on the region of the installation, how the module is oriented toward the sun, and the module type. The consumer must know the expected average incident solar irradiance (I_{region}), termed *insolation*, to estimate his or her PV module requirements. NREL keeps a database of the insolation in the United States. For instance, the annual average for a flat module with southern orientation and latitude tilt can be found on the NREL Web site (Figure 5.11). Insolation is reported in units of kilowatt-hours per square meter per day (kWhr/m²/day).

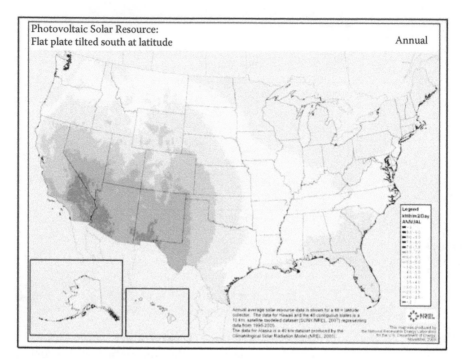

Photovoltaic Solar Resource:
Flat plate tilted south at latitude   Annual

FIGURE 5.11
National Renewable Energy Laboratory (NREL) plot of annual solar insolation in the United States for a flat module display facing south at latitude. (From NREL [National Renewable Energy Laboratory], National Center for Photovoltaics, Golden, Colorado, 2010, http://www. nrel.gov. With permission.)

The customer's power requirements ($P_{required}$) divided by the product of insolation, energy losses, the module's efficiency, and the module's area provides an estimate of the number of modules required (Equation 5.6):

$$N_{modules} = \frac{P_{required}}{\text{Loss Factor}_{DC-AC} Eff_{module} A_{module} I_{region}} \tag{5.6}$$

Example Calculation:

$P_{required}$: A consumer uses 42 kilowatt-hours per day
A_{module}: The area of the module is 826 mm × 1575 mm = 1,300,950 mm² = 1.3 square-meters
I_{region}: In Los Angeles, California, the average is 5.6 kilowatt-hours per square meter per day.
Eff_{module}: 10% per module
Loss Factor$_{DC-AC}$ conversion = 85%

$$N_{module} = \frac{(42 \text{ kWhrs/day})}{(5.6 \text{ kWhrs/m}^2/\text{day})(0.85)(1.3 \text{ m}^2)(.10/\text{module})} = 68 \text{ modules}$$

It should be noted that the module's electrical performance dictates the number of modules required. The higher the efficiency is, the fewer modules the customer will need to purchase.

The consumer's available installation area sometimes limits the number of modules that can be purchased. The dimensions of each module are included in the product sheet. The area of the module (A_{module}), measured in square meters, times the number of modules ($N_{modules}$) estimates the array's footprint ($A_{footprint}$), in square meters (Equation 5.7):

$$A_{Footprint} = A_{module}N_{modules} \tag{5.7}$$

Example Calculation:

$$A_{Footprint} = (1.3 \text{ m}^2)(68) = 88.4 \text{ m}^2$$

The calculations above are estimates. There are more robust calculators, such as PV Watts™, located at the NREL Web site. It is typical to receive 10% to 15% less power than that expected from these calculations. Variations that impact module performance but are not detailed in this example calculation include, but are not limited to, electrical wire resistance, elevated module temperature, module degradation rate, and soiling.

Differences between actual power generation and the watt peak power used for product characterization are usually the largest source of disparity between customer expectations and module performance. To decrease consumer dissatisfaction, manufacturers will report the actual power per peak module power, measured in kilowatt-hours per peak kilowatt for each year (kWh/kW$_p$-yr), for a given installation geography and conditions (Equation 5.8). Specifically, the energy yield is equivalent to the product of insolation and various performance losses (e.g., direct current [DC] to alternating current [AC] conversion, increased module temperature) divided by the test irradiance [16]:

$$\text{Energy Yield} = \frac{\text{Energy Produced}}{\text{Peak Power}}$$

$$= \frac{I_{region} \text{ Loss Factor}_{DC\text{-}AC} \text{LossFactor}_{Temp}}{\text{Test Irradiance } (1\text{kW/m}^2 \text{perASTM E1036})} \tag{5.8}$$

Superior product performance is denoted with a higher energy yield ratio, typical values are 1775 to 2400 kWh/kW$_p$-yr. The exact value is dependent on the tilt, tracking, and solar cell chemistry.

The BOS components and batteries are optional. The tracker is a principal component of the BOS for some installations. The tracker moves the module's orientation to follow the sun to ensure maximum direct sunlight. When a customer chooses not to include this in his or her purchase, the insolation value will decrease, affecting the amount of power the customer can harness each day. Similarly, batteries are also optional costs. Batteries are used to store the energy until it is required. When batteries are not included, the consumer will only benefit from PV energy when it is sunny, unless the consumer is able to sell surplus energy to the utilities for a credit against energy purchased at night or during cloudy weather.

The inverter and the installation are required purchases with the array. The inverter is required to convert the modules' power into something the consumer can use. The modules produce DC when light shines on them. However, the outlets in buildings require AC. The inverter switches the DC produced by the module to the AC required by the consumer. Certified professionals referred to the consumer by the PV manufacturer typically perform the array installation.

Consumer satisfaction of the installation affects the company's revenue. PV companies sell modules but sometimes partner with other distributors for the inverters, BOS, batteries, and installation. However, the entire system creates an impression on the consumer. Any breach in performance at the system level creates a negative impression of the PV manufacturer.

Traditionally, PV energy has found a niche in rural communities not serviced by the national grid. The cost of the infrastructure for power lines to these remote areas is higher than purchasing modules and inverters. In this scenario, the individual household owns the PV module. However, in order for PV energy to become widely acceptable for urban consumers, the cost must be significantly reduced or PV energy must replace traditional sources used by power plants.

To increase PV's competitiveness in urban environments, federal and state governments have provided incentives. Currently, there is a U.S. federal incentive program for PV consumers. It provides a 10% tax credit and 5-year accelerated depreciation for commercial installations. There are also federal mandates directed at power plant consumption. As of March 2009, the Environmental Protection Agency (EPA) reported 33 states have participated in the Renewable Portfolio Standard (RPS), a mandate that sets state-directed goals for the percentage of renewable energy that must be used by electric utilities. Each state provides an incentive program based on the local energy prices, the percentage of the population connected to the power grid, and the residents' attitudes toward green technology. Due to the high utility costs, green initiatives, and expansive land for agriculture, the PV incentives in California and Texas are some of the most generous in the United States. Using Texas as an example, there are a wide variety of tax incentives and rebates offered to residential consumers. Installation rebates are based on the number of watts purchased, and to qualify, the purchased array must have a minimum power output. Rate schedules can be found at the relevant State Energy Commission's Web site. All the details of these incentives are contained in the state-run Emerging Renewables Program.

The marginal cost (*MC*) is the cost associated with including one additional unit of output. It is the slope of the graph of total cost (*TC*) versus quantity (*Q*) produced (Equation 5.9):

$$MC = \frac{dTC}{dQ} \tag{5.9}$$

Lower marginal costs create higher profits. In order to reduce marginal costs, the total cost must be minimized.

Average total costs (*ATC*) are equal to the total costs divided by the quantity produced. Average total costs are divided into two contributions: the average fixed costs (*AFC*) and the average variable costs (*AVC*) (Equation 5.10):

$$ATC = AFC + AVC \tag{5.10}$$

Average fixed costs are the total fixed costs divided by the quantity. Fixed costs include bonded indebtedness, rent, insurance, and equipment depreciation. Most of these fixed costs are self-explanatory, except equipment depreciation.

Manufacturing equipment depreciates in value after it is first purchased. The depreciation rate is dependent on the number of units produced. The profit made on each unit decreases the loan debit incurred when the equipment was purchased. Therefore, as quantity increases, the average fixed costs decrease principally because the fixed costs are divided by larger quantities. A steady decrease will occur, but eventually a plateau is reached.

Average variable costs are the total variable cost divided by quantity produced. Contributions to total variable costs include materials, power, transportation, and labor. Average variable costs decrease with increasing quantity because both the numerator decreases and the denominator increases (Figure 5.12). For instance, bulk pricing for materials requires high volume manufacturing. As a PV manufacturer buys more polymers to build PV modules, the PV manufacturer will typically receive a lower price from the polymer manufacturer. As the price for polymers decreases and the quantity of produced PV modules increases, the average variable cost decreases. The company is exhibiting the law of increasing returns. Each added unit of variable cost is efficiently used to increase output. However, the company eventually starts to exhibit the law of decreasing returns, causing the average variable cost to turn upward. During this period, each additional unit of variable cost does not increase the factory output as much as it did when the factory was in its infancy. For instance, because some resources are fixed (e.g., capital equipment), there will be a point at which workers, a variable resource, will begin to bump into each other, decreasing their contribution to the factory's productivity.

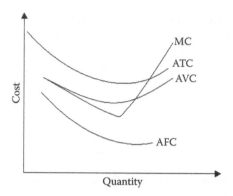

FIGURE 5.12
Average total, variable, average fixed, and marginal costs for a company.

A decrease in average total costs with an increase in production is termed *economies of scale*. Economists point to the low production volumes of PV companies to suggest they have not reached the efficiencies associated with economies of scale. However, as demand for PV power increases, each company will produce more products, decreasing their average total costs.

In addition to increasing output, a number of companies are looking for additional methods to further decrease costs. For instance, a decrease in labor costs can be achieved with the use of offshore material suppliers and manufacturing sites. In 2008, only 7% of PV module manufacturing occurred in the United States, while Japan (22%) and Europe (31%) had the highest majority [17]. Manufacturing in Asian countries is also quickly growing. This was demonstrated when BP Solar announced in March of 2010 they would be closing a plant in Maryland and moving some of those manufacturing lines to China [18].

5.5 Photovoltaic Markets Abroad

During the first energy crisis in 1977, the United States spent $65 million on renewable energy research. The entire European Economic Community spent the equivalent of $9 million between 1974 and 1979 [19]. Despite this previous economic commitment, only 3% of the energy in the United States came from renewable sources in 2009. Coal, oil, nuclear, and natural gas have a stronghold on U.S. energy consumption (Figure 5.13) [20]. Part of this devotion to carbon-based energy is the price. In 2005, the average cost of PV energy was $3.50/watt [21]. This is well above the $1 to $0.3/watt targeted by the United States Research and Development Administration in the 1970s [19]. Even with the ongoing, industry-wide cost reductions, PV is projected to remain

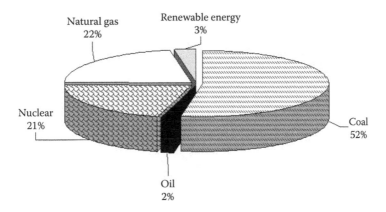

FIGURE 5.13
Energy consumption breakdown for the United States in 2009. (Data from D. Talbot, Lifeline for Renewable Power, *MIT Technology Review*, 112: 40–47, 2009.)

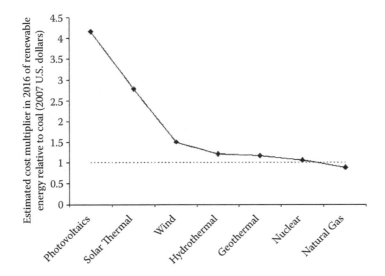

FIGURE 5.14
Projected cost ratios for various energy sources relative to coal in 2016. (Data from K. Bullis, Intelligent Electricity, *MIT Technology Review*, 112: 92–93, 2009.)

4 to 4.5 times more expensive than coal for the next decade (Figure 5.14) [22]. At these prices, there is no incentive for the U.S. consumer to switch to PV. Therefore, companies are looking abroad to the European and Asian markets to find consumers.

Renewable energy is most attractive in global regions with the highest energy consumption. High demand is likely to create energy diversification in these countries. From 1980 to 1990, the regions of highest energy demand were North America and Europe (Figure 5.15). By the mid-1990s, Asia became

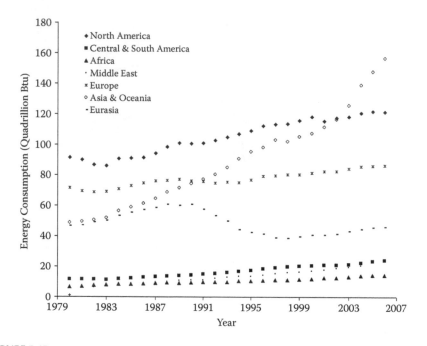

FIGURE 5.15
Energy consumption in British thermal units (1 Btu = 1.06 kilojoules) for North America, Central and South America, Africa, Middle East, Europe, Asia and Oceania, and Eurasia between 1980 and 2006. (Data from Energy Information Administration (EIA), E.1 World Primary Energy Consumption (Btu), 1980–2006, U.S. Department of Energy, Washington, DC, 2008.)

the second-largest consumer and surpassed North American consumption in the recent decade. Emerging markets are more pronounced when viewed by country. By the late-1990s, China's exponential population growth created an exponential demand (Figure 5.16). This growth is the largest contribution to this region's energy requirements.

Solar energy is the smallest sector of renewable energy for global consumption. It represents 2% of renewable energy demand, while biomass (26%), wind (34%), and hydropower (38%) almost equally share a third (Figure 5.17) [23]. However, those countries with the highest energy costs have embraced PV energy over the recent decades. The Japanese and European markets had the most explosive demand and supply of PV energy in 1990 to 2000. PV has become competitive in these countries because of the high price of traditional energy sources (Figure 5.18). The cost of heavy industrial and light industrial oil is higher in the Federal Republic of Germany, Spain, and Japan than in most parts of the United States [24]. This, in turn, makes the average power price per household larger in the aforementioned countries and grid parity a more achievable goal [25].

In 2008, the highest number of U.S. PV exports were to the Federal Republic of Germany (43%), Spain (23%), and Italy (11%) [26]. The U.S. PV

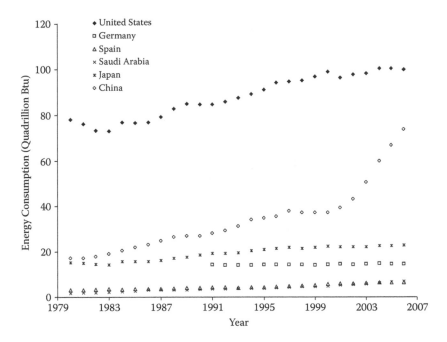

FIGURE 5.16
Energy consumption in British thermal units (1 Btu = 1.06 kilojoules) for the United States, Germany, Spain, Saudi Arabia, Japan, and China between 1980 and 2006. (Data from Energy Information Administration (EIA), E.1 World Primary Energy Consumption (Btu), 1980–2006, U.S. Department of Energy, Washington, DC, 2008.)

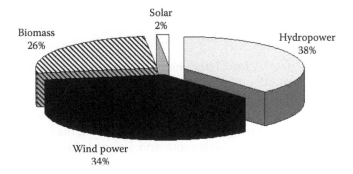

FIGURE 5.17
Energy consumption breakdown for the world in 2005. (Data from F. Ciccarello, The Role of Renewable Energy in China, Graduate thesis, Bocconi University, Milan, Italy, 2007.)

companies openly acknowledge their growth requires increasing global demand. As part of their stockholder's report issued on March 1, 2010, First Solar (Tempe, Arizona) expects the PV demand to almost double the existing demand (65 GW), including an introduction of new markets in India, China, United States, Australia, and the Arab states (105 GW) [27]. The

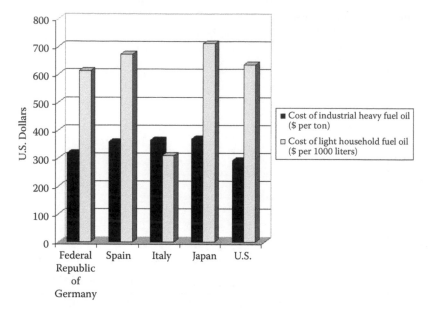

FIGURE 5.18
The cost of industrial and light household fuel oil in 2009 for the Federal Republic of Germany, Spain, Italy, Japan, and the United States. (Data from International Energy Agency [IEA], "Key World Energy Statistics," Paris, France, 2009.)

largest barrier to profitability for U.S. PV companies in these markets is competition. Competition comes from other energy sources and other global PV companies.

5.5.1 China's Solar Market

Unlike the United States, China has constantly passed legislation to promote renewable energy growth over the last five decades. Most importantly, while the United States underwent policy reversals in the 1980s, industrialization in China was forcing alternative energy initiatives to the priority of their national agenda. It was during this time that a growing manufacturing sector created higher energy requirements and larger initiatives to decrease environmental pollution.

As early as 2005, China's building sector accounted for 23% of its total energy consumption, and it was only expected to grow over the next decade [28]. During this same time period, the primary use for solar energy in China was water heating for off-grid applications in rural areas. Therefore, there has been an immediate effort to move this technology into urban on-grid applications. The results of those efforts were demonstrated during the 2008 Summer Olympics. In particular, with the Water Cube and Bird's Nest Stadiums, the Chinese have made great engineering strides in transferring energy-efficient technology to the urban populous and building facades. In addition, the seven main stadiums and the Olympic village were powered with solar energy. Similar to the United States, the integration of solar into the urban populous will remain a national focus.

In 2006, the United States barely consumed more energy than China (Figure 5.16) [29]. With the fastest-growing population in the world, increasing energy demands give China a strong motivation to become energy independent. Their policies have been organized into a three-tiered approach. The first involves public relation campaigns to gain national support from the general populous. The second tier includes policies and goals set primarily for the conversion of energy in rural environments to renewable sources. The third includes policies to support the growth of the energy market, which include tax incentives, loans, grants, and energy quotas [30]. By 2009, China's national renewable energy commitment was 8% of their total energy consumption, a larger commitment than in the United States (Figure 5.19) [31].

China's commitment to growth has made them a strong competitor in the global solar market. China created a 148% growth in renewable energy investment between 2004 and 2009. As a result, in 2009, before all of the money from the American Recovery and Reinvestment Act had been invested, China's investment of $34.6 billion for renewable energy eclipsed the United States' $18.6 billion investment [32]. The primary sectors targeted for growth in China are wind and solar energy.

Due to this investment, PV cell manufacturing has exponentially ramped over the recent years, surpassing U.S. capacity. In 2008, at least three Chinese manufacturers ranked among the top ten PV cell producers in the World [33].

There continues to be disparity between the cell efficiencies of Chinese manufacturers and other global competitors, but there is little difference in

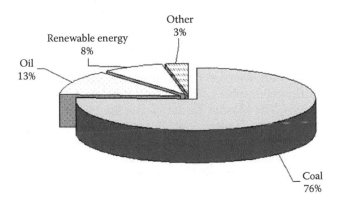

FIGURE 5.19
Energy consumption breakdown for China in 2009. (Data from J. Chen, W. Czorneck, E. Copua, M. Julian, K. Koo, and D. Zaviyalor, Renewable Energy in China: A Necessity, Not an Alternative, Wharton School of the University of Pennsylvania, 2009, http://knowledge.wharton.upenn.edu/.)

module efficiency. The Chinese dominance in semiconductor manufacturing for integrated circuits has poised its manufacturing sector for a seamless technological transfer to PV cell fabrication. In effect, Chinese companies emphasize research and development (R&D) efforts on the cell efficiency improvements and deemphasize BOS research and design. Most manufacturers have a fully automated front end and utilize manual labor for the back end [34]. Therefore, package reliability and innovation remain a potential competitive advantage for other module manufacturers.

5.5.2 Saudi Arabia's Solar Market

About half of the world's oil reserves are in the Middle East. Recognizing this opportunity to control the world's prices, the Organization of the Petroleum Exporting Countries (OPEC) was formed in 1965. OPEC is a cartel of oil-producing countries, which collude to control the world's supply of oil and therefore command a premium market price. As of 2008, it included Algeria, Angola, Ecuador, Iran, Iraq, Kuwait, Libya, Nigeria, Qatar, Saudi Arabia, the United Arab Emirates, and Venezuela.

In 1973, OPEC received the world's attention when they constrained their oil supply and increased prices. For years, Saudi Arabia has been the dominant cartel member with the largest oil reserves. They have campaigned for moderate pricing to decrease the attractiveness of alternative energy sources. This was a common point of disagreement between OPEC members. Specifically, the majority of the members wanted to increase prices for short-term gain. This dissension is cited as a reason for the unraveling of the cartel in 1985 and 1986, leading to a decrease in oil prices [4].

In 2005, as part of the requirements set forth in the U.S. Energy Policy Act, the U.S. Department of Energy prepared a report on the effect of the U.S.'s foreign oil dependency on its national security. The report listed Saudi Arabia as a country with the largest oil reserves and the strictest regulations against foreign investments [35]. Their foreign policies ensure their oil reserves were protected from current and future foreign procurement.

In addition to vast oil reserves, the Middle East has the highest solar irradiance of all world geographies. This makes it an attractive test bed for PV energy installations. The Masdar Initiative is a $15 billion investment by the city of Abu Dhabi in the United Arab Emirates to build a self-sustaining, zero-emission, zero-waste city utilizing various green technologies, most notably, solar photovoltaics and solar thermal.

The Saudi Arabian government is currently open to foreign investments and renewable energy technology transfer. There is tax exemption to attract foreign corporate investment in the Masdar Initiative. Likewise, the city contains the Masdar Institute. Its campus is staffed with world-recognized scholars to ensure the lessons learned during the city's planning and implementation are captured in test reports and academic literature [36]. The oil companies, as well as the federal government, have been investing in alternative energy as a future method of energy diversification. As evidenced from Saudi Arabia's past government policies, this market will not remain open to foreign investments forever.

5.6 The U.S. Polymer Market

In recent decades, PV companies have experienced an inversion in their variable cost structure. In 1978, the front end dominated variable costs for many companies. For instance, processing polycrystalline silicon (35%) and its conversion into cells (26%) constituted 61% of the costs. It was projected that by 1986, the back end had become the largest contribution to variable costs. The polymeric encapsulation materials alone would constitute 63% of typical manufacturing costs [37]. As predicted, today, the back end is the highest variable cost for most PV companies.

The PV industry's high encapsulation costs are the result of the types and amounts of polymers they use. This can be illustrated by contrasting their material choices against the choices made by the food and pharmaceutical industries. Ionomers, epoxies, ethylene vinyl acetate (EVA) copolymers, fluorinated polymers, and silicone rubbers are all used in food, pharmaceutical, and PV packaging (Table 5.4). However, the quantity used for each application is different. In the packaging industry, only ionomers and EVA are used in large quantities as barrier films and foams (Table 5.4). Epoxies, fluorinated polymers, and silicone rubbers are used in small quantities as adhesives, coatings, and formulation additives. These high-value, low-volume applications have evolved because of the polymer's high manufacturing costs. Following suit, the PV industry has used ionomers and EVA in high volumes as encapsulants. However, some PV manufacturers also use epoxies, silicones, and fluorinated polymers as major packaging components. In order to further reduce packaging costs, PV manufacturers will need to find alternatives to these high-cost polymers.

In Chapter 1, various classifications of polymers were introduced. Relevant classifications for PV packaging included thermoplastics, thermosets, ionomers, and elastomers. Within each of these classifications, polymers can be further subdivided into commodity, engineering, and specialty polymers. Using thermoplastics as an example, polyethylenes are commodity polymers, polyvinyl acetates are engineering polymers, and fluorinated polyolefins are specialty polymers. This distinction is based on the cost and volume of production.

The low cost of commodity polymers is the result of high-supply and low-cost monomers. Commodity resins were designed to be disposable replacements to the paper and glass used in food and pharmaceutical packaging. They are meant to have limited mechanical strength but provide an environmental barrier for the encased product. Their ease of use has made them

TABLE 5.4

Uses of Various Polymers in the Food, the Pharmaceutical, and the Photovoltaic (PV) Industries

Type of Polymer	Application in Food and Pharmaceutical Packaging	Application in PV
Ionomers	• Barrier films for skin and blister packages • Heat-sealed layer for food and pharmaceutical packaging	• Encapsulant for solar cells
Epoxy	• Surface coating • Adhesives • Primer coatings for tin-free steel beverage cans	• Adhesives • Pottants in junction boxes
Ethylene vinyl acetate (EVA) copolymer	• Packaging foam • Heat-sealed layer for food and pharmaceutical packaging	• Encapsulant for solar cells
Fluorinated polymers	• Used for strip and blister packages in pharmaceuticals	• Backsheets
Silicone rubber	• Mold release agents	• Pottants in junction boxes • Encapsulant for solar cells in selected applications

the highest-volume production of all polymer classifications since the early 1960s. The three largest thermoplastic resins based on the world's production volumes are polyethylene, polypropylene, and polyvinyl chloride. The largest suppliers of these three compounds are petrochemical companies, such as ExxonMobil and Total Petrochemicals, who creates the monomers for the polymerization as part of their refinement process during gasoline production. In 2006, 329 million barrels or 4.57% of the total U.S. annual oil production was used for feedstock in polymer manufacturing. As a basis of comparison, for that same year, the U.S. polymer industry used only 3 million barrels in energy consumption [38].

Engineering resins represent polymers designed to replace metals. The polymer structure and additives are formulated for strength, toughness, and creep resistance in load-bearing applications. They are typically composed of modified hydrocarbon chains. Polyethylene terephthalate contains an ester linkage (R_1-COO-R_2) in its hydrocarbon chain. Similarly, EVA is a copolymer that contains an acetate ester on its side chain. These structural groups require more expensive starting materials and additional chemical processing relative to commodity polymers. Engineering resins are manufactured by various chemical companies, such as DuPont (Wilmington, Delaware) and Eastman Chemical Company (Kingsport, Tennessee).

Specialty polymers are more expensive than engineering resins. The polymer chains are composed of highly engineered monomers. Therefore, as previously explained, most industries restrict the quantity of these polymers because of their higher costs. They are commonly used as coatings and additives but only make up a major portion of the product in highly engineered applications that require high lubricity, and thermal or electrical stability.

Since 1967, the majority of R&D expenditures in polymer science have come from industrial investments. By 1993, U.S. industries spent three times more on polymer R&D than the U.S. government. However, with this dominance comes control over the country's intellectual competitive advantage. U.S. industries were interested in short-term profits and incremental product improvements. This focus slowly eroded the U.S.'s technological dominance on polymer innovations. Reports issued by the National Research Council (NRC) in the early 1990s cited half of the companies filing for polymer patents in 1991 were headquartered outside the United States. This same report voices concern with increasing R&D in Germany and Japan on specialty polymers. In 1992, both countries spent 3% of their annual gross domestic product (GDP) on polymer R&D, while the United States spent 2%. Of particular concern was the increased foreign interest in research on silicones and fluoropolymers, two of the polymer resins used extensively in PV packaging [39].

Today, polymers are the third largest industry in the United States [40]. Even so, the U.S. capacity accounts for a few million of the billion pounds of global annual production. Although the exact value varies, this is typically less than a quarter of the world's production. In 2008, the top five polymer exporting markets were Canada (22%), Mexico (21%), China (8%), Belgium (4%), and Japan (4%) [40].

Between the years of 1999 and 2008, there has been a U.S. trade surplus in plastic resins production and a deficit in polymer products. Therefore, the United States is producing more resin than its manufacturing sectors require. This suggests plastic finished goods are assembled overseas and shipped back to the United States for distribution. Similarly, many PV module manufacturers are finding it more economical to perform their PV packaging overseas. To illustrate this point, DuPont's Photovoltaics division reported their Asian sales shifted from 31% in 2006 to 43% in 2008. Their sales in the Asian market almost equate with those of the European market (47%) and have completely eclipsed the North American market (10%) [41].

References

1. Chapin, D.M.; Fuller, C.S.; Pearson, G.L. May 1954. A New Silicon p-n Junction Photocell for Converting Solar Radiation into Electrical Power. *Journal of Applied Physics* 25: 676–677.
2. National Aeronautics and Space Administration (NASA). 2010. Aeronautics and Astronautics Chronology, 1958. http://www.nasa.gov

3. Harmon, C. March 2000. Experience Curves of Photovoltaic Technology. Interim Report IR-00-014, International Institute for Applied System Analysis, Laxemburg, Austria.

4. McConnell, C.R.; Brue, S.T. 1990. *Economics: Principles, Problems, and Policies,* 11th Ed. New York: McGraw-Hill.

5. Hubbard, H.M. 1989. Photovoltaics Today and Tomorrow. *Science* 244: 297–304.

6. Rotman, D. 2009. Chasing the Sun. *MIT Technology Review* 114: 44–51.

7. Narum, D. 1992. A Troublesome Legacy: The Reagan Administration's Conservation and Renewable Energy Policy. *Energy Policy* 20: 40–53.

8. Rotman, David. 2009. Can Technology Save the Economy? *MIT Technology Review* 114: 44–52.

9. Takahashi, D. 2008. Solar Boom. *MIT Technology Review* 111: 30.

10. International Science Panel on Renewable Energies (ISPRE). 2009. Research and Development on Renewable Energy: A Global Report on Photovoltaic and Wind Energy. www.icsu.org

11. National Renewable Energy Laboratory (NREL). 2010. Best Research-Cell Efficiencies. www.nrel.gov

12. Bowden, M.J. 2004. Moore's Law and the Technology S-Curve. *Current Issues in Technology Management* 8: 4–7.

13. Aravantinos, E.; Fallah, M.H. 2006. A Methodology to Improve the Mobile Diffusion Forecasting: The Case of Greece. Proceedings of ISPIM 2006 Conference Networks for Innovation, Athens, Greece, June 11–14.

14. de Tarde, G. 1903. *The Laws of Imitation.* New York: H. Holt and Company.

15. Kazmerski, L.L. October 25–28, 2004. Photovoltaics R&D: At the Tipping Point. 2004 DOE Solar Energy Technologies Program Review Meeting, Denver, CO. www.nrel.gov

16. Nishikawa, W.; Horne, S.; Melina, J. 2008. LCOE for Concentrating Photovoltaics (CPV) 2008 International Conference on Solar Concentrators for the Generation of Electricity (ICSC-5), November 16–19, Palm Desert, CA.

17. Navigant Consulting. 2009. Photovoltaic Manufacturer Shipments, Capacity and Competitive Analysis, 2008/2009.

18. Dishneau, D. May 24, 2010. BP Solar Ends Md. Manufacturing, Cutting 320 Jobs. www.siliconvalley.com

19. Pulfrey, L.D. 1978. *Photovoltaic Generation.* New York: Van Nostrand Reinhold.

20. Talbot, D. 2009. Lifeline for Renewable Power. *MIT Technology Review* 112: 40–47.

21. Henderson, R.M.; Conkling, J.; Roberts, S. 2007. SunPower: Focused on the Future of Solar Power. www.mitsloan.mit.edu/MSTIR/sustainability/SunPower/Pages/default.aspx

22. Bullis, K. 2009. Intelligent Electricity. *MIT Technology Review* 112: 92–93.

23. Ciccarello, F. 2007. The Role of Renewable Energy in China. Graduate thesis, Bocconi University, Milan, Italy.

24. International Energy Agency. 2009. Key World Energy Statistics. www.iea.org

25. Lorenz, P.; Pinner, D.; Seitz, T. June 2008. The Economics of Solar Power. Energy, Resources, Materials. *The McKinsey Quarterly.*

26. U.S. Energy Information Administration (EIA). 2010. Table 3.14: Destination of US Photovoltaic Cell and Module Export Shipments by Country, 2007 and 2008, Form EIA-63B, Annual Photovoltaic Module/Cell Manufacturers Survey. www.eia.doe.gov

27. First Solar. March 1, 2010. First Solar Corporate Overview. http://phx.corporate-ir.net/
28. Chang, J.; Leung, Y.C.; Wu, C.Z.; Yuan, Z.H. 2005. A Review on the Energy Production, Consumption, and Prospect of Renewable Energy in China. *Renewable Energy* 30: 1973–1988.
29. U.S. Energy Information Administration (EIA). 2008. E.1 World Primary Energy Consumption (Btu), 1980–2006. www.eia.doe.gov
30. National Renewable Energy Laboratory (NREL). 2004. Renewable Energy Policy in China: Overview. NREL/FS-710-35786. www.nrel.gov
31. Chen, J.; Czorneck, W.; Copua, E.; Julian, M.; Koo, K.; Zaviyalor, D. 2009. Renewable Energy in China: A necessity, Not an Alternative. Wharton School of the University of Pennsylvania. http://knowledge.wharton.upenn.edu/
32. The PEW Charitable Trusts. 2010. Who's Winning the Clean Energy Race? Growth, Competition and Opportunity in the World's Largest Economies. Washington, DC.
33. International Energy Agency. December 2008. Newsletter of the IEA Photovoltaic Power Systems Programme. www.iea.org
34. Marigo, N. 2007. The Chinese Silicon Photovoltaic Industry and Market: A Critical Review of Trends and Outlook. *Progress in Photovoltaics: Research and Applications* 15: 143–162.
35. U.S. Department of Energy. 2006. Energy Policy Act 2005: Section 1837: National Security Review of International Energy Requirements. www.pi.energy.gov
36. Bullis, K. 2009. A Zero-Emissions City in the Desert. *MIT Technology Review* 112: 56–63.
37. Sittig, M. 1979. Solar Cells for Photovoltaic Generation of Electricity Materials, Devices and Applications. *Energy Technology Review*. Park Ridge, NJ: Noyes Data Corporation.
38. U.S. Energy Information Administration. 2010. Question: How Much Oil Is Used to Make Plastic? http://tonto.eia.doe.gov/ask/crudeoil_faqs.asp
39. National Research Council Committee on Polymer Science and Engineering. 1994. *Polymer Science and Engineering: The Shifting Research Frontiers*. Washington, DC: National Academies Press.
40. Society of the Plastics Industry. 2010. SPI Plastics Industry Facts. http//spi.cms-plus.com/files/industry/plastics_industry_facts.pdf
41. Fu, Li Guang. 2009. DuPont (China) Photovoltaic Technology Center. www2.dupont.com

6

Other Polymeric Applications in Photovoltaic Modules

6.1 Emerging Polymeric Applications

Because photovoltaic (PV) cells will ultimately reach a technological limit, scientists have investigated alternative packaging materials to improve efficiencies. Antisoiling, antiscratch, and antireflective coatings can be externally applied to PV modules to increase the concentration of incident light and offset diminished performance due to environmental exposure. This is a growing field for commercial chemistries, and a number of PV manufacturers are eager to integrate antisoiling or antireflective technologies into their products.

Module improvements can also be made by substituting traditional packaging components with new materials. High index of refraction polymers can be used as encapsulants to improve PV module performance. These polymers maximize optical coupling between the glass and underlying PV cell. However, due to their higher costs, the benefit rarely warrants this material selection over commodity encapsulants.

6.1.1 Soiling Behavior of Photovoltaic Modules

Dirt, dust, soot, and animal excrement can build up on module's surfaces hindering light transmission and, subsequently, the module's performance. PV modules are typically elevated on roofs or trackers out of daily human traffic. As a result, they are not easily accessed for cleaning. Soiling decreases the transmission over all wavelengths, but the highest amount of scattering occurs in the visible region.

Researchers reported reproducibility issues for the performance measurements of modules in the same locations and between similar climates. Despite these inconsistencies, it is generally accepted that soiling has a significant impact on module performance, and it is common to experience a 3% to 4% decrease in power within 2 years of installation. A double-digit decrease in power is common for prolonged soiling of flat modules.

Field testing typically occurs for 24 months when evaluating new antisoiling technologies. This extensive evaluation procedure is required because most antisoiling samples will temporarily perform worse than uncoated substrates. Although this phenomenon is not understood, it necessitates data be averaged over an extended time period to get an accurate prediction of performance. Furthermore, it is not uncommon for glass to have a grace period lasting between 1 and 3 months before a change in transmission is observed due to soiling [1]. Due to this lengthy qualification process, researchers have looked for artificial methods to verify the desired performance.

6.1.1.1 Considerations for Developing a Soiling Protocol

It is a priority for PV manufacturers to predict performance degradation from laboratory measurements rather than assume the expense of field testing in multiple environments. PV manufacturers want to simulate climates with the highest number of PV installations. These include polar, temperate, arid, and tropical climates. Each of these climates has a characteristic soil type and annual rainfall accumulation.

Soil differs in each location due to climate, indigenous organisms, topography, and degree of erosion. National databases and the United Nations (UN) compile soil types for all world geographies. Specifically, the U.S. Department of Agriculture (USDA) developed a taxonomy to classify soil based on chemical and physical features. The orders of taxonomy with greatest PV significance are alfisols, mollisols, aridisols, entisols, oxisols, and ultisols (Table 6.1). In temperate climates, soils are defined by high concentrations of organic material from forests, termed *alfisols*, or grasslands, termed *mollisols* (Table 6.2). Arid climates typically contain *aridisols* and *entisols*, rich in humus and finely pulverized minerals. Finally, tropical climates contain red dirt and clay, called *oxisols* and *utilsols*, which receive their coloration from high iron oxide content.

Due to the difficulty in duplicating this complicated chemistry, many researchers purchase commercially formulated artificial soils to eliminate experimental variability. Commercial soils are homogeneous with defined particulate sizes. Aridisols are the easiest to purchase because they have

TABLE 6.1

Climates, Representative Countries, and Common Soil Types for Areas with the Highest Solar Investments and Installations

Climate	Country	Soil Type
Temperate	United States	Ultisols, mollisols, aridisols, alfisols, inceptisols
Temperate	Germany	Alfisols, inceptisols
Arid	Saudi Arabia	Aridisols, entisols
Temperate	China	Ultisols, inceptisols

TABLE 6.2

Selected U.S. Department of Agriculture (USDA) Soil Orders and Descriptions

Soil Order	Description
Ultisols	High concentration of minerals and clays; red, orange, or yellow soil containing high concentrations of iron oxide with a slightly acidic pH
Oxisols	Red or yellowish soil with high iron oxide and aluminum oxide concentrations; small amounts of clay and organic matter from the surrounding tropical rain forests
Mollisols	High organic content and nutrient rich
Aridisols	Pulverized rock, minerals, and humus
Alfisols	Decomposed forest cover and organisms with enriched clay
Entisols	Mainly unaltered parental rock

been used for decades to certify filter performance in automobiles and refrigerators. Powder Technology, Inc. (Burnsville, Minnesota) and Particle Technology Labs (Downers Grove, Illinois) both sell test soils with the chemical formulation and particulate sizes required for test standards used in the automotive and the appliance industries.

The USDA's taxonomy does not distinguish between soils based on process of origin or elemental composition. However, the term *soil* can generically refer to all chemical compounds formed from both natural and man-made processes (Table 6.3). These soils can be further subclassified based on elemental composition. Organic compounds contain carbon atoms. These include biological materials and polymeric decomposition products from incineration, landfill, mechanical abrasion, or weathering. Inorganic compounds are formed from metallic atoms. Pulverized rocks and minerals, from natural rock erosion, and some components of diesel soot, from automobiles and factories, are examples of inorganic soils.

Within each climate, soiling will differ based on population density (i.e., rural versus urban). Average power loss for a module in an urban environment is twice that in rural [2]. This increased soiling is caused by a higher concentration of aerosols in urban environments. Aerosols are a gaseous blend of organic and inorganic solids and liquids. They are the vehicles for dirt accumulation on modules. The composition of aerosols has been deformulated to typically include sand, moisture, salt, fungal spores, pollen, nitrogen oxide (NO_x), and sulfur oxide (SO_x). The majority of the particulates are minerals (65%), with a smaller portion composed of combustion and biological by-products.

The combustion by-products of industrialization are considered to be the most deleterious compounds for polymeric degradation. The reactive nature of these compounds makes them an initiator in decomposition reactions. Different polymers have different susceptibility to those chemicals. For instance, polyesters and polyacetals are particularly susceptible to hydrolysis caused by environmental acids created from soot and rain.

TABLE 6.3

Selected Organic and Inorganic Soils: Their Source and Examples

Chemical Type of Soil	Source	Examples
Organic	Natural	Bacteria, fungus, fingerprints
	Man-made	Soot (carbon and acid emissions)
Inorganic	Natural	Dust
	Man-made	Diesel soot (NO_x, SO_x)

There have been multiple proposed soiling theories, but the theory proposed by Cuddihy and Willis is the most relevant to PV packaging. They theorized a three-layered soiling structure formed through a combination of chemical and physical processes. Each layer requires a different washing process to be removed [3].

The first layer is chemisorbed to the glass surface, referring to a chemical reaction that covalently bonds the soil particulate to the glass. It is hypothesized that this reaction is preceded by a functionalization of the surface due to acid rain; however, a complete chemical mechanism for these reactions has not yet been reported. Furthermore, an extension of this hypothesis has not been proposed to describe soiling in low soot, agricultural environments.

The second layer has strong physisorption to the first layer. Physisorption requires mutual attraction between partial charges on the soiling particulate and the material substrate. The attraction between the particles and the surface is a result of weak attraction between Van der Waals or dipolar forces. The strength of the attraction gradually decreases with decreasing soil depth until the top layer is simply physically packed particulates. The largest particulate diameters are always closest to the top.

The third layer is loosely settled soil. There is a light mechanical stacking of the particulates, and they can easily be removed with wind, minimal mechanical abrasion, or rain.

In contrast, mechanical soiling theories require surfaces to have physical features that allow soil particulates to become embedded in the superstrate. This is a particular concern when glass surfaces are textured to increase light transmission. Textured glass is prone to higher soiling rates than smooth, featureless glass. The dirt and microscopic features mechanically interlock, requiring aggressive mechanical scrubbing to remove the soil.

The installation configuration can also influence soiling behavior. A number of researchers have commented on the correlation between incline angle and soiling rate. Most report an exponential decay in light transmission eventually reaching a performance plateau. For example, Hegazy and coworkers placed glass at varying angles from 0° to 90° on a roof in the Minia region of Egypt for 1 month. They monitored changes in the transmittance dust factor (F_d) defined as the ratio of the transmission through a soiled substrate (τ_d) versus the transmission through a clean substrate (τ_c). They found that regardless of the angle, the data fit a decaying exponential curve, where D represents the days of exposure, and a, b, and c are empirical constants. The rate of decrease is defined by the exponential term, and the plateau is described by $(1 - a)$ (Equation 6.1):

$$F_d = \frac{\tau_d}{\tau_c} = (1-a) + a\exp(bD^c) \tag{6.1}$$

The largest, most rapid decreases ($a = 2.193$, $b = -0.0221$, $c = 0.522$) occurred with a completely horizontal incline, and the smallest, slowest ($a = 0.0495$, $b = -0.118$, $c = 0.635$) decreases occurred with a completely vertical incline. The vertical orientations contained mostly small particulates less than 1 micron in diameter. Although this observation has been documented in a number of studies, the reason is not completely understood. It is hypothesized that the small particles become chemically adhered to the surface by a tie layer of dew that forms overnight. The working theory suggests this thin layer of water changes the surface hydrophilicity, increasing soil absorption [4], and the gravitational force is not large enough to overcome the particulate's adhesion to the surface.

Garg and coworkers found polymeric substrates exhibit larger decreases in transmission than glass, regardless of the installation incline and environment [5]. Rarely can unsoiled transmission be restored after washing the soiled, polymeric surface. Based on this observation, it is hypothesized that a chemical attraction occurs between most polymeric substrates and soiling particulates.

Both the polymeric chains and additives can influence physisorption on polymeric surfaces. Full and partial charges on the chains cause the polymer to exhibit higher soiling tendencies. For instance, ionomers and polyesters exhibit higher soiling due to their full and partial charges, respectively. These effects can be countered with various additives. As an example, carbon black decreases the electrostatic charge buildup on the polymeric surface, decreasing dust and dirt attraction. Of course, carbon black cannot be used in areas of modules that require optical transparency. Therefore, transparent polymeric components require specialized antisoiling technology and periodic rainfall to maintain high transmission.

The National Oceanic and Atmospheric Administration (NOAA) and various manufacturers of weatherometers keep an up-to-date measurement of the world's precipitation. Relevant climates in descending order, according to annual rainfall, are tropical, temperate, and arid. A number of outdoor studies have revealed there is rarely a change in module performance during light rainfall. Therefore, a minimum total accumulation is required to identify a performance change. However, an experimental simulation should not exceed the total annual accumulation anticipated for the climate of product deployment. For reference, annual mean rainfall in Singapore, China (1°22′N 103°59′E, elevation 15 m), a tropical climate, is 2300 millimeters (mm). In Dhahran, Saudi Arabia (26°32′N 50°13′E, elevation 92 m), an arid climate, the annual mean rainfall is 88 mm [6].

Most manufacturers agree the highest amount of soiling is expected in desert and tropical climates. In the desert, water is a commodity that cannot be spared for cleaning. PV manufacturers would like to integrate antisoiling technology onto their modules that will repel dirt and minimize accumulation. Tropical environments are a close second because of the correlation between high humidity and accumulation. Each night a new formation of dew attracts a new layer of soil resulting in a multilayered soil construction that is difficult to remove without mechanical scrubbing.

6.1.1.2 Experiments to Characterize Antisoiling Coatings

Scientists try to predict soiling by characterizing a surface's wettability. Wettability describes how a liquid spreads across the surface. The contact angle formed between the substrate, liquid, and air interfaces is a standard wettability metric. When water is used as a dispensing fluid, the contact angle can be used to measure hydrophilicity. An image of a water droplet on the substrate is taken immediately after it is dispensed to avoid evaporation. The measured angle between the substrate and the droplet is the contact angle. Angles higher than 90° define a hydrophobic surface, and those with a contact angle below 90° are called hydrophilic. The same measurement performed with oil provides a measurement of oleophilicity.

Young's equation defines the contact angle (θ) as a balance between creation forces. Surface tension is the force applied per unit area, measured in newtons per square meter (N/m^2), required to create surface area between two phases (Figure 6.1). The cosine of the contact angle is equivalent to the differences between interfacial tension of the vapor–solid ($\gamma_{VS} = \gamma_s$) and liquid–solid (γ_{LS}) phases divided by the interfacial tension at the liquid–vapor ($\gamma_{LV} = \gamma_L$) phase (Equation 6.2):

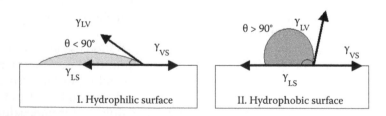

FIGURE 6.1
The equilibrium surface tensions that determine a contact angle for a water droplet on a (I) hydrophilic surface and (II) hydrophobic surface.

$$cos\theta = \frac{\gamma_{VS} - \gamma_{LS}}{\gamma_{LV}} \tag{6.2}$$

A low contact angle indicates the liquid–solid tension is favored due to lower creation force. Conversely, a high contact angle indicates the solid–vapor phase is favored.

The surface chemistry, as defined by surface energy, allows researchers to predict wettability. Surface energy is a combination of dispersion and polar energies of attraction between the liquid and surface molecules. The dispersive energy occurs between all surfaces and liquids, but polar energy requires polar molecules in both phases. The Good and Van Oss model can be used to measure the surface energy defined as a combination of dispersive (γ_S^d) and polar interactions (γ_S^+, γ_S^-) (Equation 6.3):

$$\gamma_S = \gamma_S^d + 2(\gamma_S^+ \gamma_S^-)^{1/2} \tag{6.3}$$

The surface energy can be derived from the experimental contact angle (θ) using three different liquids with known energy (γ_L), dispersive interactions (γ_L^d), and polar interactions (γ_L^+, γ_L^-), available in various material property handbooks [7] (Equation 6.4):

$$\gamma_L(1 + cos\theta) = 2((\gamma_S^d \gamma_L^d)^{1/2} + (\gamma_S^+ \gamma_L^-)^{1/2} + (\gamma_L^+ \gamma_S^-)^{1/2}) \tag{6.4}$$

Three equations and three unknowns (γ_S^d, γ_S^+, γ_S^-) allow the experimenter to derive the surface energy (γ_S) for the new coating.

A high surface energy signifies a strong attraction between the surface and liquid, causing good wettability and a low contact angle. A low surface energy corresponds to a weak interaction between the liquid and solid molecules, creating poor wettability and a high contact angle. The reader should be aware that a number of other equations have been proposed based on the substrate chemistry. However, the Good and Van Oss Model is the most applicable to PV packaging materials.

There is confusion in the PV industry as to whether a hydrophobic or a hydrophilic surface is required to minimize module soiling. It has been empirically observed that a layer of soil decreases the glass' contact angle. Because chemically similar surfaces attract, there is a concern that hydrophilic glass will attract more dirt. However, some believe that a hydrophilic surface will allow the water to evenly coat the glass, pulling the dirt off as it slides off the module. Conversely, hydrophobic surfaces may attract less hydrophilic soil. Those particles that do adhere may be pulled into the droplet as it rolls off the module. However, the soil must be in the path of the water droplet to be pulled off the hydrophobic glass.

Because of the wide variety in soil chemistries, one coating chemistry is unlikely to perform well in all installations. Many material manufacturers provide both a hydrophilic and a hydrophobic antisoiling technology so that PV manufacturers can choose the appropriate material for their installation site.

There are a number of commercial instruments that can be helpful for material qualification and maintenance protocol development of PV arrays. Gardner, Erichsen, and Sheen washability testers are instruments used in a number of industries to develop a maintenance protocol. Complementary ASTM testing standards, D4488 [8], D2486 [9], and D4828 [10], provide a testing procedure for these instruments. These instruments are abrasion testers with a detergent added between the brush and the test substrate. After the completion of the user-defined cycles, performance is evaluated based on the manufacturer's criteria. For PV manufacturers, ultraviolet-visible (UV-Vis) transmission measurements are typically used to verify there is no change in desired optical properties. Modules can also be constructed from the washed substrates, and changes in short circuit current (I_{sc}) can be correlated with decreased PV cell illumination due to remnant soil or scratches on the glass.

6.1.1.3 Antisoiling Coatings

Even during the PV industry's infancy, soiling was viewed as a large technical obstacle for profitability. In the early 1980s, the Jet Propulsion Laboratory (JPL) began surveying materials as potential candidates for antisoiling coatings. The only formulations that showed promise were developmental formulations supplied by 3M (St. Paul, Minnesota) and Dow Corning (Midland, Michigan).

Willis reported the experimental results of these developmental formulations in the Ninth Annual JPL report as part of the program for terrestrial PV development. Glass (Sunadex™), acrylic (Acrylar™), and fluoropolymers (Tedlar®) were the test substrates. Each was coated with either Dow Corning's developmental formulation E-3820-103B, a perfluorodecanoic acid attached to the surface with a silane coupling agent, or 3M's L-1668, an undisclosed fluorinated silane. The substrates were assembled over a PV cell and placed outdoors in Enfield, Connecticut, for 48 months. For all substrates, Dow Corning's E-3820-103B worked better than 3M's L-1668 formulation, and a rainfall was required for both coatings to provide any benefit [3].

The soiling behavior of the uncoated substrates indicated acrylics and fluoropolymers are more prone to soil collection. The short circuit current (I_{sc}) decreased for PV cells behind acrylic (–9%) or a fluoropolymer (–4.7%). The current of cells behind uncoated glass increased (0.8%) (Table 6.4). This further confirmed the industry's requirement that the superstrate material must have a high mechanical modulus to avoid diminished performance due to soiling.

After soiling, the modules coated with E-3820-103B had smaller decreases in short circuit current (I_{sc}) than uncoated modules. The decrease was smallest for glass (–0.7%), indicating the antisoiling chemistry was more effective for nonpolymeric substrates (–1.9% for acrylic and –2.6% for a fluoropolymer).

The high cost of fluorinated chemistry has hindered its use by the PV industry. In the 1980s, a number of American chemical companies decreased

TABLE 6.4

Observed Changes in Short Circuit Current (I_{sc}) after 24 Months of Soiling for Glass, Acrylics, and Fluoropolymers with and without a Fluorinated Coating, Dow Corning's E-3820-103B

Substrate	Percent Change in I_{sc} for Uncoated Substrates	Percent Change in I_{sc} for E-3820-103B Coated Substrates
Glass	0.8%	–0.7%
Acrylic	–9.0%	–1.9%
Fluoropolymer	–4.7%	–2.6%

Source: Data from P.B. Willis, "Investigation of Test Methods, Material Properties and Processes for Solar Cell Encapsulants," Annual Report, ERDA/JPL-954527, Springborn Laboratories, Inc., Enfield, Connecticut, July 1977.

funding for these solar research and development efforts. Therefore, Dow Corning's E-3820-103B was never commercialized. However, Dow Corning currently sells fluorosilane chemistry for electronic displays. Its stability during outdoor use is unknown.

Although failing to find a commercial formulation for immediate product integration, Willis did conclude efficacy required a hard, hydrophobic, smooth, and low surface energy coating.

Easy-to-clean polymeric surfaces have been a desired property in the flooring industry since the late 1960s. Polyurethane, epoxy, and acrylic coatings have been adopted by that industry [11]. These coating have the same qualities specified by JPL, but these polymers have not been successfully transferred to commercial PV applications. However, most PV companies are less interested in easy-to-clean surfaces and more interested in self-cleaning technologies.

Most self-cleaning technologies have been developed around the use of anatase crystallized titanium dioxide. Titanium dioxide (TiO_2) exhibits a photocatalytic response. When placed on a glass superstrate, the TiO_2 can decompose organic soil that falls on its surface. Impinging light forces electrons in the TiO_2 crystalline structure from the valance band to the conduction band. This light must have equivalent energy to the TiO_2 band gap ($E_g = 3.18$ eV) to create an electron-hole pair. The hole (h^+) can react with water creating a hydroxyl radical ($^•OH$), while the electron (e^-) reacts with oxygen to create a superoxide ion ($O_2^{-•}$) (Figure 6.2). These two highly reactive species undergo a series of free radical reactions with organic soil to decompose it into carbon dioxide and water.

Dip and spray coatings are the most popular application techniques for this chemistry. In both cases, polymers are part of the carrier solvent. For instance, Negishi and coworkers dip coated soda lime glass in a suspension of polyethylene glycol and 2-(2-ethoxy ethoxy)ethanol and then sacrificed the polymer, leaving behind a nanoporous TiO_2 structure [12]. Spray coating of TiO_2 has been commercialized for use on building facades and windows.

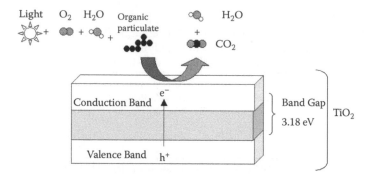

FIGURE 6.2
Self-cleaning reactions for anatase titanium dioxide (TiO_2).

Various commercial TiO_2 formulations are also available, such as NuTiO™ from Bio Shield, Inc. (Port Richey, Florida) and SolarStucco™ from Green Earth Nano Science, Inc. (Toronto, Canada). Again, polymers act both as a solvent and a sacrificial carrier in these chemistries, and they are not typically present during the coating's service.

The Lotus Effect is another natural chemistry identified as self-cleaning. Scientists have been trying to mimic this behavior since the early 1960s. However, it was not until a decade later, after the invention of scanning electron microscopy (SEM), that it was possible to visualize the cleaning phenomena as a combination of microstructure and chemistry. The lotus leaf surface is characterized by papillae 10 to 15 microns in width and 10 to 20 microns in height. A wax on the leaves increases surface hydrophobicity, causing polar molecules to contract to limit their interfacial contact (Figure 6.3). All large, hydrophilic particulates will sit on top of the papillae. Incident water will create a contact angle on the order of 160°, giving rise to the term *superhydrophobic* when describing the lotus leaf surface. When inclined, the water will roll off the surface, gathering soil as it rolls off the leaf.

Since the mid-1990s, chemical and material scientists have been trying to commercially duplicate the Lotus Effect. One of those chemistries specially developed for glass substrates is BASF's (Florham Park, New Jersey) Lotus Effect aerosol spray. It contains nanoparticles that self-assemble into an ordered structure while waxes and short chains of polyethylene and polypropylene create a hydrophobic surface.

Although these self-cleaning technologies have been marketed for external use, they have not been specifically validated for PV applications. Unlike building facades, the PV module performance is directly linked to the coating durability and efficacy. More specifically, the coating must not discolor, haze, or induce light scattering during weathering. Therefore, ultimately, PV manufacturers will need to verify a coating selection with a cost-benefit analysis and durability measurement. Most manufacturers target an upper limit of 5 to 10 U.S. cents per module for both material and processing costs. Of course,

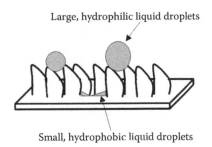

Large, hydrophilic liquid droplets

Small, hydrophobic liquid droplets

FIGURE 6.3
A large, hydrophilic droplet sitting on papillae, and a small, hydrophobic droplet between them.

this cost target is dependent on the longevity of the benefit. The industrial expectation is that a coating will maintain the same level of efficacy for 25 to 30 years. The author is unaware of a chemical manufacturer that has put forth a formulation that simultaneously adheres to all these requirements.

6.1.2 Antiscratch Coatings

A PV module's performance and aesthetics can be altered with scratches. For instance, when thermoplastic frames are scratched, they are not only aesthetically unpleasing, but their electrical resistance can be compromised if the thermoplastic is an insulator for electrical conduits. This same design struggle has occurred in a number of industries. In this instance, it is helpful to leverage the techniques and experiences of the automotive industry to develop testing procedures to screen engineering polymers for PV frames [13].

Because the device and experimental parameters may be tailored to the manufacturer's needs, scratch and indentation tests are commonly used techniques. A scratch tester includes a pin with a weight attached to the head. The pin is dragged along the surface to create a trenched profile. The scratch is analyzed with a microscope and compared to a pristine surface to identify the depth of penetration. The lower the penetration depth, the better are the antiscratch properties. A second indirect measurement for scratch resistance is hardness. Microindentors are a common instrument for measuring this property. A probe with a known hardness and area is inserted into the surface of the coating with a known force. The hardness of the substrate is calculated as the maximum applied force divided by the penetration area. Softer materials will have a lower hardness given the same amount of force. However, there is an inverse relationship between scratch resistance and hardness. Specifically, a substrate with a higher hardness is more prone to scratching.

Thermoplastics and thermoplastic elastomers have been used extensively in the automotive industry for car interiors (e.g., dashboards) and car exteriors (e.g., bumpers). The commodity polymers, principally polypropylenes, have been heavily integrated due to their ease of processing and low cost. Inorganic fillers are typically added to increase impact strength and decrease cost. However, those reinforced grades are prone to deep scratches due to regions of lower plastic deformation. To avoid the natural tendency of polypropylene to scratch, Ciba (Basel, Switzerland) chemical company developed new additives, sold under the trade name Irgasurf® SR 100. With a 3 wt% addition of Irgasurf to polypropylene formulations, the polymer's antiscratch properties have become equivalent to engineering polymers.

The superstrate glass of a PV module must also be scratch resistant to avoid optical loss. Only polysiloxane hardcoats have the optical clarity and weathering characteristics required for coating module glass. Polysiloxane hardcoats can be applied with spray, dip, or flow coating. Despite their ease of application, high material costs cause polysiloxanes to be cost prohibitive. Because the effects of scratches on the module performance have not been widely characterized, a cost-benefit analysis is difficult to perform. Therefore, PV manufacturers are trying to combine this attribute with other high-value properties such as antisoiling and antireflectivity.

6.1.3 Antireflective Coatings

Antireflective (AR) coatings are used to decrease the reflection of light between two interfaces with different refractive indices. They are tailored to minimize reflectance at specific regions of the electromagnetic spectrum, such as infrared (IR), visible, or ultraviolet (UV) light. PV manufacturers would like to apply an AR coating to the superstrate glass of their modules, because the air–glass interface is one of the largest areas of index mismatch in the assembly.

The two most common AR coatings for PV applications are one- or two-layer interference chemistries designed to minimize reflectance in the visible region. A single layer of inorganic coating, with an intermediate refractive index between air and glass, is applied in a quarter-wavelength thickness to impart AR properties (Figure 6.4). Both the gradient in refractive index and the controlled coating thickness force more transmitted light through the glass. The ideal refractive index is the geometric mean of the two original interfaces, for PV modules that is the refractive index of air (n_{air}) and the refractive index of glass (n_{glass}) (Equation 6.5):

$$n_{AR} = \sqrt{n_{air}n_{glass}} \tag{6.5}$$

In addition, the quarter-wavelength thickness causes destructive interference between light of equal intensity, reflected from the air–coating and coating–glass interfaces. Historically, magnesium fluoride has been the most common chemistry used in single-layer coatings. Further improvements in AR can be achieved with alternating layers of refractive index. Two layers allow for a smaller step change in refractive index across the interface. In addition, two

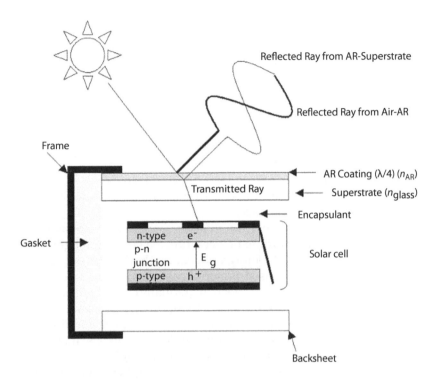

FIGURE 6.4
An impinging light ray on a single-layer antireflective coating.

quarter-wavelength thicknesses further increase destructive interference at each interface. These two-layer coatings are typically a combination of silica and titanium dioxide applied with dip coating, an application well suited to the high-volume production requirements of PV manufacturing.

Polymer laminates used in the electronic industry as antismudge, antireflective, and antiscratch coatings for personnel electronics are the one exception to inorganic AR coatings. Vikuiti™, distributed by 3M, is a laminate construction of an antifingerprint layer, antireflective coating, matte hardcoat, polyethylene terephthalate, adhesive, and disposable liner. These have been successful for indoor applications, but they must certainly be reformulated for outdoor use. Currently, there is no outdoor equivalent offered by 3M.

6.1.4 High Index of Refraction Polymers

Looking at the cross section of a module, there is a large disparity between the refractive index of PV cells (e.g., silicon) and packaging materials (e.g., glass and encapsulant) (Table 6.5). For instance, soda lime glass is about two units away from silicon. Most PV manufacturers make packaging choices to minimize the refractive index change across the glass–encapsulant interface.

TABLE 6.5

Packaging Function with Corresponding Inorganic or
Organic Materials and Refractive Index

Function	Materials	Refractive Index
Superstrate	Optical glass	1.6
Encapsulant	Polydimethylsiloxane	1.4
High refractive index encapsulant	Polyphenylene vinylene	2.1
Solar cell	Silicon	3.5

They then use an AR coating on the cell to maximize light transmission through the encapsulant–cell interface.

This same variation occurs across the interfaces of electronic assemblies, such as waveguides, optical recording devices, tunable lasers, and light-emitting diodes. In those industries, AR inorganic coatings have fallen out of favor due to difficult processing, increased material costs, and lower mechanical toughness. Instead, their approach has been to gradually change the refractive index across the interfaces to minimize loss. This approach for solar manufacturers requires polymeric encapsulants with a range of refractive indices.

Most commercial polymers have a refractive index close to one. Increasing that value closer to inorganic substances requires changing the chemical structure to a highly conjugated chain, such as polyphenylene vinylene. There has also been some industrial interest in integrating inorganic chemistry into the polymer chain. This approach, of high refractive index encapsulants, has not found a place in current PV manufacturing because they are cost prohibitive.

6.2 Concentrated and Organic Photovoltaics

Between 2004 and 2008, silicon supply constraints caused more than a doubling of raw material cost [14]. In an effort to further reduce production costs, PV manufacturers have tried to reduce the amount of costly semiconductor material used in the module. These initiatives have generated interest in concentrated photovoltaics (CPVs).

CPVs use polymeric lenses and packaging components to focus incident light on encapsulated cells, thereby increasing collection efficiency over a smaller cell area. Concentrated photovoltaic modules constituted 125 peak kilowatts, 0.06% of U.S. PV exports, in 2005, and 27,527 peak kilowatts, 2.8% of U.S. PV exports, in 2008 [15]. This explosive growth is the outcome of increased competitiveness due to improvements in cell efficiency. CPV has the highest cell efficiency, ~28% relative to other technologies, such as

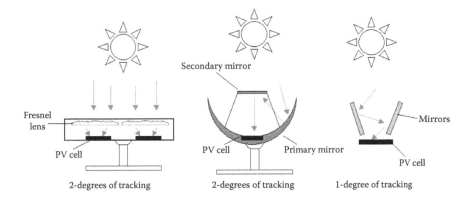

FIGURE 6.5
Depiction of a lens, parabolic mirrors, and reflectors with corresponding tracking requirements.

single crystalline silicon (~25%) and thin-film technologies (~13% to 20%) (Figure 5.4).

PV concentrators use either lenses, mirrors, or a combination to magnify incident irradiance (Figure 6.5). The concentrator type is categorized based on geometric configuration, concentration factor, and tracking requirements.

6.2.1 Lenses

There are a number of design considerations to ensure a lens efficiently focuses light rays onto a focal point. The two relevant for this discussion are geometry and material selection. Design considerations include the geometric shape, the wavelength with the highest quantum efficiency for the PV cell, and the refractive index of the materials used for the lens.

Fresnel lenses are the most common geometry designed for CPV concentrators [16]. A Fresnel lens is composed of a number of Fresnel zones visualized as a series of prisms with different steps in thickness cut around the lens circumference. The expected concentration factors in the assembly are typically modeled using ray tracing, specifically the edge-ray principle [17]. The edge-ray principle solely considers the trajectory of edge rays from the source through the lens and to the target. This modeling is available in a number of optical design programs, such as ZEMAX®. The concentration factor typically ranges from 5 to 500× depending on if the lens is planar or circular. A planar Fresnel lens utilizes one-axis tracking, and a circular requires dual-axes tracking.

The material requirements for polymeric lenses are based on optical and thermal properties. The material class must have high optical transmission. In Chapter 2, the wavelength dependency of polymeric refractive indices was discussed. As discussed, the material choice is dependent on the wavelengths that need to be focused. For PV cells, the relevant wavelengths are always in the visible spectrum. This requires a limited dispersion of visible wavelengths of light, as measured by the Abbe number. The Abbe number (V_D) defines how much the refractive index changes over the visible spectrum. It is the ratio of the difference between the refractive index of green-yellow (n_D = λ = 589 nm) light minus 1 and the difference between the refractive index at blue-green (n_F = λ = 486 nm) and red (n_C = λ = 656 nm) light.

$$V_D = \frac{(n_D - 1)}{(n_F - n_C)} \tag{6.6}$$

In addition, the optical design is sensitive to the polymeric lens thickness and the radius of curvature. As a result, the polymeric material must have a low coefficient of thermal expansion (CTE) to ensure the focus point does not change during thermal cycling.

Historically, polyacrylates have been the favored material selection for Fresnel lenses due to their high optical transmission, low wavelength dispersibility (Abbe number 50 to 60), and comparatively good weathering characteristics [18]. Because concentrators are most efficient under sunny conditions, most of the weathering of polymethylmethacrylate (PMMA) lens has been performed in arid environments. Rainhart and coworkers found a 10% drop in transmission over a 17-year exposure period and significant decreases in mechanical strength due to crazing [19]. Reports issued by 3M indicate similar decreases for optical transmission after 13 years of weathering in Minnesota with the largest decreases occurring between 350 and 500 nm.

These decreases in performance were due to poor mechanical and soiling durability. Polymethylmethacrylates are susceptible to crazing, in part, because they are amorphous with an above ambient glass transition temperature. Crazes are small microvoids on the polymer surface created by thermal or mechanical stresses. They severely decrease the mechanical strength of the polymer and readily form cracks under applied stress. Rainhart and coworkers noted there was a significant increase in embrittlement due to crazing in weathered samples. The samples also suffered mechanical scratches during soiling. This accounted for a 7% optical transmission loss [19]. To avoid this decreased performance, abrasion-resistant polyacrylates have been proposed for these applications. Abrasion-resistant polyacrylates are formulated with a cross-linked polysiloxane topcoat; a commercial example is Spartech's (Clayton, Missouri) Polycast SAR.

6.2.2 Metallic Films

Parabolic concentrators and reflectors use metallic surfaces to concentrate light. Parabolic concentrators are composed of two parabolic mirrors, referred to as the first and the secondary. The first mirror reflects light from the Sun to the secondary mirror elevated above the first. The secondary mirror focuses light back onto the underlying PV cell. Parabolic concentrators often exhibit high concentration factors, and they require a dual-axes tracker to follow the course of the Sun and optimize performance. In contrast, reflectors have guidance mirrors on each side of the PV cell. Light rays from the Sun hit the mirror and bounce down to the cell. The mirrors are at a fixed position, limiting the concentration ratios to low values of less than 5× Suns when utilizing a single axis of tracking.

Polymers have been proposed as alternative mirror glazing for concentrators and reflectors since the 1970s; however, they have failed to become widely commercialized due to various technical limitations. The primary focus of the glazings is to act as an environmental barrier to protect the underlying metal, typically silver. In addition, the glazing must be optically clear, impact resistant, weather resistant, and inexpensive to manufacture. Polycarbonate, polyacrylates, and polyesters were surveyed by the industry in the 1970s. Only polyacrylates exhibited no significant degradation after 2 to 5 years of outdoor exposure in Arizona.

Silvered polymer reflectors are a laminate including a heat-sealable polymer film, a tie-layer adhesive, metal foil, and a protective barrier film (Figure 6.6). Their reduced cost, ease of manufacturing, and mechanical flexibility make laminates an attractive alternative to silver-plated glass. Their

FIGURE 6.6
A polymeric, metallic film.

commercialization has been limited by higher than expected production costs, poor optical weathering, and delamination in moisture and thermal cycling.

In the mid-1990s, 3M manufactured ECP-305+, a polyacrylate evaporated with silver, protected with copper, and adhered with a pressure-sensitive adhesive. They also had a separate product line, sold as SS-95, consisting of silver evaporated polyester with a thin protective coating of polyacrylate. The SS-95 film was quickly discontinued due to large optical losses in field tests. Specifically, the National Renewable Energy Laboratory (NREL) confirmed a 30% reduction in optical properties of SS-95 after less than 5 years of outdoor exposure in Colorado. In contrast, ECP-305+ film demonstrated less than 5% reflection loss after 10 years of outdoor exposure at NREL facilities but suffered delamination in the field. The product was discontinued due to high manufacturing costs, insufficient demand, and poor consumer image [20].

NREL has published the most recent weathering data on glazings and oversees reflector development for the Department of Energy (DOE). Through joint development with commercial vendors, they have been able to extend the life of polymeric mirrors. NREL and ReflecTech® jointly developed an undisclosed formulation that was recently marketed to PV manufacturers. The first attempts significantly weathered under 7 to 8× Suns. The film's hemispherical reflectance decreased to approximately 70% from 95% in less than two UV equivalent years of simulated weathering. In 2005, after further material modifications, additional samples were submitted for testing. Currently, the material is rated for 10 years based on Arizona outdoor weathering. Despite these improvements, neither this nor any other commercial product has been warranted for the target 25 to 30 years.

6.2.3 Luminescent Solar Concentrators (LSCs)

The goal of luminescent solar concentrators (LSC) is to simultaneously decrease costs and increase efficiency. Traditionally, LSCs are designed to absorb unusable light and re-emit it at wavelengths with the highest efficiency for the underlying PV cell. By increasing the concentration of highest-efficiency light, material costs decrease because fewer PV cells are required for the same power generation. Unlike the aforementioned concentrator techniques, there is no required tracker creating odd-shaped array footprints in residential areas.

LSCs are constructed of a polymeric lens that directs the light into the adjacent PV cells. The flat plate geometry is the most widely discussed, but cylindrical concepts have been proposed and patented since the late 1970s [21–24]. Both geometries are constructed from a polymer, typically PMMA. The light enters the largest face, bounces internally, and exits the edges. The concentration factor (G_{geom}) is the geometric ratio of the area of the face (A_{face}) to the area of the edge (A_{edge}) (Equation 6.7, Figure 6.7).

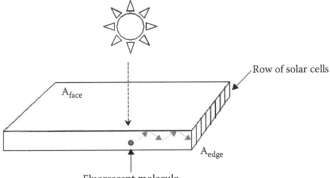

FIGURE 6.7
A polymeric luminescent solar concentrator (LSC) with a single dye molecule absorbing and emitting light toward a line of adjacent solar cells.

$$G_{geom} = \frac{A_{face}}{A_{edge}} \qquad (6.7)$$

For a rectangular geometry, 3 mm thick and 1 m long, the concentration factor would be 333×. This theoretical value is not realized in practice. Currently, the highest reported value is a 4% efficiency improvement [17]. In order to become competitive, LSCs must demonstrate a 6% to 10% increase in efficiencies above those of the solar cell. These poor practical demonstrations are due to a number of chemical and optical limitations.

The polymeric LSC relies on a fluorescent dye to convert the incident light into usable wavelengths. These dye molecules must absorb low-wavelength, high-energy light and emit high-wavelength, low-energy light. Each molecule purchased from a chemical manufacturer has a characteristic absorption and emission curve, and both are commonly represented as a Bell curve of intensity versus emitted or absorbed wavelengths. The difference between absorption and emission maxima is defined as the Stoke's shift.

This is the same phenomena discussed for UV quenchers in Chapter 2. Therefore, some polymer formulations will have this inherent property built into the commercial formulation. Polymers with these UV additives will fluoresce green. However, the additive concentration is typically not high enough to exhibit LSC behavior. Therefore, PV manufacturers will have to request a specific fluorescent additive, with the desired characteristics, be added to the polymer. When choosing a fluorescent molecule, the absorption curve of the encapsulant's formulation (e.g., UV stabilizers) should not overlap the fluorescent molecule's absorption or emission curves in order to optimize conversion efficiency.

Similar to photovoltaic cells, fluorescent molecules are characterized by a quantum efficiency curve. The quantum efficiency of the dye is how much of the absorbed light is converted to emitted light. A quantum efficiency of 100% is ideal, but commonly unrealized. Most commercial organic dyes have an efficiency of 75% to 80%.

At high loadings, adjacent dye molecules molecularly couple, decreasing conversion efficiency. This means they absorb the light emitted from surrounding dye molecules rather than the impinging light, thereby decreasing the efficiency of the conversion of incident light [25]. As an example, Kurian and coworkers demonstrated a 24% decrease in quantum yield (75% to 51%) for a 6× increase in fluorescent rhodamine 6G dye concentration in PMMA ($1.5 \cdot 10^{-4}$ to $9 \cdot 10^{-4}$ mol/l) [26]. Therefore, it is essential to create a homogenous dye distribution at the lowest concentration possible to avoid this conversion inefficiency.

The response of the PV cell and dye pair must be optimized. The emission curve must overlap with the internal quantum efficiency (IQE) of the PV cell, but the absorption curve must not rob useful wavelengths from the cell. Coumarin-6 overlapped with a single crystalline silicon cell IQE response is a commonly proposed pair (Figure 6.8). The absorption peak maximum occurs at 450 nm where the cell is 68% efficient and emitted at 500 nm where the cell is 71% efficient. Ideally, the absorption maximum would be between 350 and 400 nm where the cell is less than 50% efficient and emitted between 550 and 600 nm where the cell is 80% efficient. Unfortunately, this means the

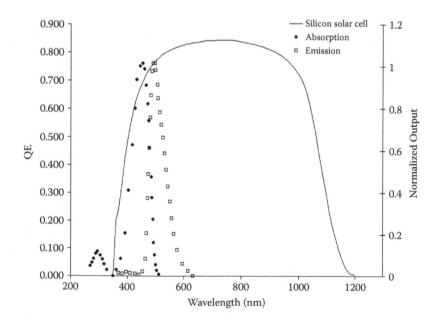

FIGURE 6.8
Single crystalline silicon solar cell internal quantum efficiency (IQE) curve overlapped with an absorption and emission curve for coumarin-6.

desired absorption curve is further than a typical Stoke's shift away from the desired area of emission.

In response, researchers have proposed a double-down conversion approach, also known as a two-dye system. In this case, two fluorescent dye molecules are incorporated into the LSC. The absorption curve of one dye has an emission curve that overlaps with the absorption of another. The emission of the second dye molecule overlaps with the IQE of the PV cell. Of course, the limitation is that both dyes must have high quantum efficiencies and the emitted light must perfectly couple inside the LSC to be fully effective.

Fluorescent chemistry has been extensively used in the textile and toy industries for decades; however, these organic molecules have short lifetimes. Some glow-in-the-dark shirts and toys contain fluorescein, an organic molecule that has an absorption peak maximum of 494 nm and emission peak maximum at 521 nm. Most of these commercially available formulations have low stability because they were formulated for commercial goods with short lifetimes. The fluorescent molecules photoxidize under prolonged exposure to UV light and oxygen. Photoxidation results in a loss of fluorescent properties and a color shift, also known as lightfastness or photobleaching. This remains a poorly understood and predicted mechanism. The dye industry typically does not guarantee color stability for more than 2 to 10 years of outdoor exposure.

When LSCs were first proposed in the late 1970s, Batchelder and coworkers performed stability measurements on laser dyes, rhodamine-6G tetrafluoroborate and coumarin-6. The dyes were impregnated into PMMA plaques at relatively low concentration (10^{-4} moles/liter) and exposed to light and dark cycles at 60°C and 50°C, respectively. The UV source was a fluorescent bulb, and the chamber was kept at 100% Relative Humidity (RH). Photobleaching was monitored by absorption measurements at various exposure intervals. The behavior was characterized by a rapid decrease followed by a plateau, modeled as a decaying exponential. These dyes were projected to last for 2 to 6.8 equivalent years. Admittedly, Batchelder did not identify the chemical mechanism for degradation, or its dependence on moisture or oxygen ingress [23]. The dye's short lifetime indicates the device efficiency would exponentially degrade during the 25- to 30-year power warranty offered by PV manufacturers. It is important to note these academic studies are not performed on fully packaged modules. Therefore, a PV manufacturer needs to perform his or her own testing to verify these aging characteristics.

In order for LSCs to maximize efficiency, the converted light must be coupled into the adjacent PV cell. To achieve coupling, there must be a number of internal reflections before the light escapes from the LSC into the PV cell. Based on Snell's law, the refractive index multiplied by the angle of incidence is equal to the refractive index of the second medium and the angle of transmittance through that medium (Equation 6.8):

$$n_1 \sin \theta_1 = n_2 \sin \theta_2 \qquad (6.8)$$

There will be internal reflections when the emitted light is less than the critical angle. To find the critical angle (θ_c), the exiting angle of light must be 90° to the polymeric interface ($\theta_2 = 90°$). This makes the critical angle equivalent to the inverse sine of the ratio of the refractive indices of the two media (Equations 6.9 and 6.10):

$$n_1 \sin\theta_c = n_2 \sin 90°$$ (6.9)

$$\theta_c = \sin^{-1}\left(\frac{n_2}{n_1}\right)$$ (6.10)

If the surrounding medium is air, then the equation for internal reflection is solely dependent on the polymer matrix (n_1) (Equation 6.11):

$$\theta_c = \sin^{-1}\left(\frac{1}{n_1}\right)$$ (6.11)

The difficulty is a fluorescent dye molecule in the center of the concentrator will emit light in 360°. Therefore, the majority of light will be lost as it exceeds the critical angle of escape. A number of researchers have modeled this loss in flat PMMA plates to find approximately 26% of the emitted light from the point source is lost due to escape from the LSC [23,27].

In order to increase the internal reflections, researchers proposed a series of additional processing layers laminated to the surface of the LSC. These additional layers would contain liquid crystals that would transmit all wavelengths but selectively reflect those emitted wavelengths of highest efficiency for the PV cell. When included in the LSC structure, they are referred to as photo-band stop filters. In addition to this approach, a number of researchers have investigated processing methods to allow the dye molecules to preferentially align, increasing the concentration of emitted light trapped at the critical angle. Unfortunately, each of these new materials and processing steps increase production costs, making LSCs a less competitive alternative to the aforementioned CPV techniques.

6.2.4 Polymeric Photovoltaic Solar Cells

Entire books have been devoted to the niche discipline of polymeric photovoltaic solar cells, also known as organic solar cells. It is not the intent of this section to cover the expansive research in this area but to provide the reader with an overview of the application's current feasibility and limitations as it relates to polymeric packaging.

Like CPV, polymeric photovoltaics are a cost-reduction technology. Specifically, the manufacturing costs can be reduced by eliminating inorganic chemistry, sourced from precious Earth metals, and using polymers, mainly sourced from petroleum by-products.

Polymeric substrates are flexible and can be formed into a number of different geometries, increasing the commercial applications for PV. If polymer packaging is used, the modules are also flexible. Polymeric photovoltaics have been integrated into automobile components, such as the hoods and roofs, and personal apparel, including backpacks and blankets. The former has public-sector applications, while the latter has military applications.

The use of polymers eliminates the need for Restriction of Hazardous Substances (RoHS) and Waste Electrical and Electronic Equipment (WEEE) exemption discussed in Chapter 2. Polymers can be easily formulated to exclude these restricted inorganic elements and brominated flame retardants, because these chemicals are not part of the inherent chemistry of the polymer chains. Instead, they are small molecular additives placed in the commercial formulations. Formulation of compliant materials simply requires manufacturers to avoid restricted additives. Additive substitution is a challenge for polymer manufacturers, but it is not an impossibility. In contrast, certain semiconductor chemistries used for PV cells require restricted elements in their structure. These restricted substances cannot be excluded without completely reformulating the cell chemistry.

The polymeric PV module is a multilayered structure. The specific chemistry used in each layer is proprietary, but there are some materials that have been commonly used in various academic and commercial pursuits. The superstrate cover is either glass or a transparent polymer depending on if the module needs to be rigid or flexible, respectively. The next layer is indium tin oxide (ITO), a cathode layer. A polymer insulates the cathode from the active layer, commonly composed of poly-3,4-ethylenedioxythiophene-polystyrenesulfonate (PEDOT-PSS). The photoactive substrate contains a combination of electron donors, highly conjugated polymers, and electron acceptors, dopant molecules (e.g., nanomaterials). The photoactive layer is sandwiched with another insulating layer, commonly sodium fluoride, and an anode. The anode is commonly an opaque metal (e.g., aluminum, silver, or gold).

There are a number of technical limitations for polymeric photovoltaics. Most importantly, the efficiency (2% to 7.9%) is two to five times lower than commercial inorganic formulations. This lower efficiency requires a larger surface area to get the same amount of power. Even with reduced costs, this new technology commonly cannot be competitive in high-power applications, such as residential and commercial installations.

There have been significant reliability issues due to the polymeric packaging used for polymeric solar cells. Their performance will decrease exponentially in the presence of water and oxygen. Some devices will not operate longer than a few hours when exposed to the air. In effect, polymeric solar cells have more stringent water vapor transmission rate (WVTR) and oxygen transmission rate (OTR) specifications (10^{-7} to 10^{-6} g/m^2/day, 10^{-7} to 10^{-3} cm^3/m^2/day) than food packaging (10^{-1} to 10^1 g/m^2/day, 10^{-1} to 10^1cm^3/m^2/day) and thin-film PV cell applications (10^{-4} to 10^{-3} g/m^2/day, 10^{-4} to 10^{-3} cm^3/m^2/day).

Packaging options are limited because polymer manufacturers have focused on providing packaging for inorganic PV cells, which constitute a larger segment of the photovoltaic market.

These requirements are not insurmountable; light-emitting diodes (LED) have similar requirements. Historically, LED technology has used silicones for encapsulation; however, new commercial laminate structures are a cheaper alternative. These laminates are constructed from a transparent layer of aluminum oxide (Al_2O_3) or silicon oxide (SiO_x) sandwiched between thermoplastic polymers. Rollprint sells polyethylene terephthalate–coated aluminum oxide and silicon oxide films under the trade name ClearFoil®. The addition of inorganic layers improves the permeant barrier characteristics by dropping the ingress rates by at least an order of magnitude. For example, ClearFoil® exhibits a lower WVTR and OTR (0.025 to 1.55 $g/m^2/$ day, 0.062 to 0.62 $cm^3/m^2/day$) than a laminate of polyethylene terephthalate and polyethylene (4.65 $g/m^2/day$, 69.77 $cm^3/m^2/day$) [28]. Some of these films have been commercialized specifically for organic solar cell applications. As an example, Ceramis® is a multilayered polylactic acid (PLA), silicon oxide film sold by Alcan Packaging (Asheville, North Carolina).

Even though these transmission rates are still above the specification, these improvements have translated into longer service lives when metallized foils are used for encapsulation. Lungenschmied and coworkers extended their polymeric solar cell service life from 6 hours when packaged in polyethylene terephthalate to 6000 hours when packaged with metallized polyethylene naphthalate [29]. However, how these barrier properties age due to weathering remains unknown.

Because polymeric solar cells remain in their developmental infancy, chemical and material scientists will need to develop a new class of polymers and compounds to make this a viable commercial option. In April 2010, 3M announced that they were working on a next-generation product composed of polyethylene terephthalate and polyethylene naphthalate that would specifically address the 10^{-6} $g/m^2/day$ WVTR specification [30].

References

1. Kimber, A.; Mitchell, L.; Nogradi, S.; Wenger, H. 2006. The Effect of Soiling on Large Grid-Connected Photovoltaic Systems in California and the Southwest Region of the United States. Conference Record of the 2006 IEEE 4th World Conference on Photovoltaic Energy Conversion, 2: 2391–2395.
2. Sawyer, R.F.; Pitz, W.J. September 28, 1982. Assessment of the Impact of Light Duty Diesel Vehicles on Soiling in California. Agreement A2-064-32. www.arb.ca.gov

3. Willis, P.B. July 1977. Investigation of Test Methods, Material Properties and Processes for Solar Cell Encapsulants. Annual Report, ERDA/JPL-954527, Springborn Laboratories, Inc., Enfield, Connecticut.
4. Hegazy, A.A. 2001. Effect of Dust Accumulation on Solar Transmittance through Glass Covers of Plate-Type Collectors. *Renewable Energy* 22: 525–540.
5. Garg, H.P. 1974. Effect of Dirt on Transparent Covers in Flat-Plate Solar Energy Collectors. *Solar Energy* 15: 299–302.
6. ATLAS. 2001. Weathering Testing Guidebook. Atlas Electric Devices Company, Pub No. 2062/098/200/AA/03/01.
7. Van Oss, C.J. 2006. *Interfacial Forces in Aqueous Media,* 2nd Ed. Boca Raton, FL: CRC Press.
8. ASTM D4488-95. 2008. Standard Guide for Testing Cleaning Performance of Products Intended for Use on Resilient Flooring and Washable Walls—Historical Standard ASTM International, West Conshohocken, PA, DOI: 10.1520/D2486-06, www.astm.org
9. ASTM D2486-06. 2008. Standard Test Methods for Scrub Resistance of Wall Paints, 2000 ASTM International, West Conshohocken, PA, 2008. DOI: 10.1520/D4488-95, www.astm.org
10. ASTM D4828-94. 2008. 2003 Standard Test Methods for Practical Washability of Organic Coatings, 2000 ASTM International, West Conshohocken, PA, DOI: 10.1520/D4828-94R08, www.astm.org
11. Verhorf, L.G.W. 1988. *Soiling and Cleaning of Building Facades.* Boca Raton, FL: Taylor & Francis.
12. Negishi, N.; Takeuchi, K.; Ibusuki, T. 1998. Preparation of the TiO_2 Thin Film Photocatalyst by the Dip-Coating Process. *Journal of Sol-Gel Science and Technology* 13: 691–694.
13. Martin, J.W.; Bauer, D.R. 2002. *Service Life Prediction Methodology and Metrologies.* Oxford Press.
14. Bartlett, J.E.; Margolis, R.M.; Jennings, C.E. September 2009. The Effects of the Financial Crisis on Photovoltaics: An Analysis of Changes in Market Forecasts from 2008 to 2009, NREL/TP-6A2-46713, National Renewable Energy Laboratory, Golden, Colorado.
15. U.S. Energy Information Administration (EIA). 2010. Figure 3-5. Photovoltaic Cell and Module Shipments by Type, 2004–2008, Form EIA-63, Annual Photovoltaic Module/Cell Manufacturers Survey.
16. Neil, M.J.; Piszczor, M.F.; Eskenazi, M.I.; McDanal, A.J.; George, P.J.; Botke, M.M.; Brandhorst, H.W.; Edwards, D.L.; Hoppe, D.T. August 3–8, 2003. "Ultralight stretched Fresnel lens solar concentrator for space power applications," SPIE's 48th Annual Meeting, San Diego.
17. Leutz, R.; Suzuki, A. *2001 Nonimaging Fresnel Lenses: Design and Performance of Solar Concentrators.* Berlin: Springer-Verlag Berlin Heidelberg.
18. Bäumer, S. 2005. *Handbook of Plastic Optics.* Weinheim: Wiley-VCH Verlag GmbH.
19. Rainhart, L.G.; Schimmel, W.P. 1975. Effect of Outdoor Aging on Acrylic Sheets. *Solar Energy* 17: 259–264.
20. Kennedy, C.E.; Terwilliger, K. 2005. Optical Durability of Candidate Solar Reflectors. *Transactions of the ASME* 127: 262–269.

21. Mauer, P.B.; Turecheck, G.D. "Fluorescent Solar Energy Concentrator." U.S. Patent 4 149 902 April 17, 1979.
22. Oster, E.A. "Tubular Luminescent Solar Collector-Photocell Structure." U.S. Patent 4 251 284 February 17, 1981.
23. Batchelder, J.S.; Zewail, A.H.; Cole, T. 1979. Luminescent solar concentrators. 1: Theory of operation and techniques for performance evaluation. *Applied Optics* 18: 3090–3110.
24. Bliedes, H.R.; Yerkes, J.W. "Luminescent Solar Collector," U.S. Patent 4 357 486 November 2, 1982.
25. Lee, J.-H.; Teng, C.-C.; Lin, J.-H.; Lin, T.-C.; Yang, C.C. 2005. Direct Observations of Energy Transfer and Quenching Dynamics between Alq3 and C545T in Thin Films with Different Doping Concentrations. *Proceedings of SPIE* 5632: 66–71.
26. Kurain, A.; George, N.A.; Paul, B.; Nampoori, V.P.N.; Vallabhan, C.P.G. 2002. Studies on Fluorescene Efficiency and Photodegradation of Rhodamine 6G Doped PMMA Using a Dual Beam Thermal Lens Technique. *Laser Chemistry* 20: 99–110.
27. Richards, B.S.; McIntosh, K.R. 4 to 8 Septemer 2006. "Ray-Tracing Simulations of Luminescent Solar Concentrators Containing Multiple Luminescent Species," 21st European Photovoltaic Solar Energy Conference, Dresden, Germany 185–188.
28. Dodrill, D. 2010. Advances in Clear High-Barrier Packaging. www.rollprint. com
29. Lungenschmeid, C.; Dennler, G.; Czeremuszkin, G.; Latreche, M.; Neugebauer, H.; Sariciftci, N.S. 2006. Flexible Encapsulation for Organic Solar Cells. *Proceedings of SPIE* 6197: 619712-1–619712-8.
30. Hitschmann, G. 2010. Development of High Barrier Films for Flexible PV and OLED Applications. *Printed Electronics and Photovoltaics Europe*. Dresden, Germany.

Appendix A: Conversion Factors and Common Units of Measurement

TABLE A.1

Metric System Prefixes, Symbol, and Conversion Factors

Prefix	Symbol	Conversion Factor
Giga	G	1,000,000,000
Mega	M	1,000,000
Kilo	k	1,000
Hecto	h	100
Deca	da	10
		1
Deci	d	0.1
Centi	c	0.01
Milli	m	0.001
Micro	μ	0.000001

TABLE A.2

Example of Applying These Prefixes to Length Measured in Meters

Prefix	Symbol	Conversion Factor
Gigameter	Gm	1 Gm = 1,000,000,000 meters
Megameter	Mm	1 Mm = 1,000,000 meters
Kilometer	km	1 km = 1,000 meters
Hectometer	hm	1 hm = 100 meters
Decameter	dam	1 dam = 10 meters
Meter	m	1 meter
Decimeter	dm	1 dm = 0.1 meters
Centimeter	cm	1 cm = 0.01 meters
Millimeter	mm	1 mm = 0.001 meters
Micrometer	μm	1 μm = 0.000001 meters

TABLE A.3

Common Properties, International System of Units and Abbreviations

Property	SI Units	Abbreviation
Band gap	Electron volt	eV
Concentration	Moles per liter	mol/l
Current	Ampere	A
Dielectric strength	Volts per millimeter	V/mm
Energy	Joules	J
Impact resistance	Joules per meter	J/m
Irradiance	Watts per square meter–nanometer	$W/(m^2 \bullet nm)$
Melt flow index	Grams per 10 minutes	g/10 min
Modulus	Newtons per square meter = pascal	$N/m^2 = Pa$
Molecular weight	Grams per mole	g/mol
Oxygen transmission rate	Cubic centimeter per square meter per day	$cm^3/m^2/day$
Peel strength	Newtons per meter	N/m
Power	Watts = joules per second	W = J/s
Strain	Percent	%
Stress	Newtons per square meter	N/m^2
Temperature	Kelvin	K
Thermal conductivity	Watts per meter–kelvin	$W/(m \bullet K)$
Viscosity	Newton–second per square meter	$(N \bullet s)/m^2$
Voltage	Volts	V
Volume resistivity	Ohms–centimeter	$\Omega \bullet cm$
Water vapor transmission rate	Grams per square meter per day	$g/m^2/day$

Appendix B: Glossary

Active area: The surface area of the photovoltaic module responsible for the conversion of light into electricity

Adhesive: A material used to bond two surfaces together

Amorphous silicon cell (a-Si): A classification of thin-film photovoltaic cells composed of noncrystalline silicon that lacks long-range order and uniform lattice structure; typical E_g = 1.7 eV

Antireflective coating: Typically an inorganic coating formulated to decrease the reflection and increase the transmission of specific wavelengths of light

Backsheet: A material typically composed of a polymer used as a primary barrier to the backside of a photovoltaic module

Balance of Systems (BOS): Components of the installation used to mount the array to the roof and electrically connect it to the home or business

Band gap (Eg): The energy difference between the valence and conduction bands of a semiconductor material, more commonly referred to as the light energy required for a photovoltaic module to generate energy

Cadmium telluride (CdTe): A thin-film photovoltaic cell composed of a semiconductor constructed of the elements of cadmium and tellurium; typical E_g = 1.44 eV

Casting: Common processing technique used for elastomers and thermosets; an enclosure is used to give the polymer shape, and it is removed after the cure reaction is complete

Cell degradation: Chemical and physical processes that decrease the photovoltaic cell's electrical performance over time

Cell interconnects: Electrical connections between cells

Cell string: A series of cells connected together in order to increase electrical output

Circular Fresnel lens: Visually appears similar to a planar Fresnel lens but focuses light onto a spot

Compression: A mode of mechanical deformation defined by pressing the specimen between two plates

Concentrated photovoltaics: The use of mirrors, lenses, or both to concentrate light on a small area of photovoltaic cells and generate electricity

Copper indium diselenide (CIS): A thin-film photovoltaic cell made from a semiconductor composed of the elements copper, indium, and selenium; typical E_g = 1.04 eV

Copper indium gallium diselenide (CIGS): A thin-film photovoltaic cell made from a semiconductor composed of the elements of copper, indium, gallium, and selenium; typical $E_g = 1.67$ eV

Dielectric material: Electrical insulator that creates charge separation when an electric field is applied

Dielectric strength: The minimum electric field that causes polymeric breakdown

Directional hemispherical reflectance: The amount of reflected light from a surface when only irradiated with direct light

Elastic modulus: The slope of the stress versus strain curve in the elastic deformation region

Elastomer: A classification of polymers known for their rubbery mechanical behavior and chemical cross-links

Electrical insulator: A material used to separate and stabilize electrical charges

Electromagnetic spectrum: Categorization of the various wavelengths of electromagnetic radiation, also known as the light spectrum

Encapsulant: An intermittent polymer layer used to encase the photovoltaic cells in a photovoltaic module

Engineering strain: Change in length divided by the initial length

Engineering stress: Deformation force divided by the specimen's initial area

Ethylene vinyl acetate copolymer (EVA): A copolymer composed of ethylene and vinyl acetate monomers

Fill factor: The peak maximum power divided by the power that would be generated if the device could simultaneously produce at open circuit voltage and short circuit current

Fluorinated polyolefin: A thermoplastic polymer with at least one fluorine atom in its hydrocarbon chain

Free radical: A highly reactive chemical species with an unpaired electron in its atomic orbital

Grid lines: Electrical contacts on photovoltaic cells

Hydrolytic degradation: A chemical mechanism requiring the presence of water which changes the polymer's chemical composition and physical properties

Injection molding: A method of polymer processing used to convert polymer pellets into a new shape. It involves both heat and pressure created by a reciprocating screw and requires the molten polymer to be injected into a mold carved into the customer's desired shape.

Insolation: Electromagnetic radiation on an area in a specified geographic location, time of day, and angle of orientation to the sun

Inverter: Electrical components required to covert the panel's direct current (DC) to the alternating current (AC) output required by the home or business outlets

Ionomer: A classification of polymers that have a pendant ionic group along the backbone of the polymer chain

Irradiance: The power of the radiant electromagnetic light incident on an area per wavelength of light

Junction box: A container used to house electrical wires and connections

Lamination: A processing technique used in photovoltaic packaging requiring heat and pressure to melt and encapsulate the photovoltaic cells

Lap shear test: An adhesive test that requires pulling two substrates simultaneously in opposite directions

Mechanical strength: The peak stress a specimen experiences before it breaks

Mechanical toughness: A material's resistance to fracture when under stress

Mole: A unit of measure equivalent to Avogadro's number (6.022137×10^{23})

Multijunction cell: A single solar cell composed of multiple thin-film semiconductor chemistries

Open circuit voltage: The voltage at zero current

Organic cell: A photovoltaic cell composed of polymers responsible for converting light into electricity, also known as a polymeric photovoltaic cell

Oxidative degradation: A chemical mechanism requiring the presence of oxygen that alters both the polymer's chemical composition and its physical properties

Packaging factor: The effect of the packaging on the maximum power produced by the photovoltaic cells

Peak maximum power: The maximum power produced on a voltage–current curve of a photovoltaic cell or module

Peel test: An adhesive test that requires pulling one substrate while holding the second stationary. The direction of force is either 90° or 180° relative to the stationary substrate.

Perfectly competitive market: A market described by no barriers to entry or exit, a high quantity of suppliers, and each supplier sells a homogenous product with no ability to set market prices

Photoelectric effect: The phenomenon of converting light into electricity; specifically, the active material emits electrons due to incident light rays

Photovoltaic array: A group of electrically connected photovoltaic modules

Photovoltaic cell: Smallest division of a photovoltaic array capable of generating power, typically composed of a semiconductor material that converts light into electricity

Photovoltaic module: A series of connected and encapsulated photovoltaic cells representing the smallest unit of energy generation available for consumer purchase, also known as a photovoltaic panel

Photovoltaic thermal energy: Harnessing light energy to generate thermal energy; high-temperature collectors use parabolic mirrors to concentrate light on a tube of heat transfer fluid used to power a turbine and generate electricity, also known as solar thermal

Planar Fresnel lens: A type of Fresnel lens that focuses light into a beam

Polar molecule: A molecule with a separation of partial charges into a slightly positive region and slightly negative region of the molecule

Polyacetal: A classification of thermoplastic polymers with an acetal group in the chain and synthesized from monomers with an aldehyde or ketone functional group

Polyacrylates: A classification of thermoplastic polymers known for their optical transparency and mechanical strength and synthesized from monomers with an acryl group, also called acrylics, and polymethylmethacrylate is one example

Polycarbonate: A thermoplastic polymer with carbonate groups in the polymer backbone and composed of the monomers bisphenol A and phosgene

Polycrystalline silicon photovoltaic cell: A type of crystalline silicon photovoltaic cell composed of multiple silicon crystals characterized by multiple grain boundaries

Polyene: A polymer with multiple carbon double bonds in its backbone

Polyester: A classification of thermoplastic polymers with an ester group in the chain; polyethylene terephthalate is one example

Polyethylene naphthalate: A type of polyester known for its barrier properties synthesized from the monomers ethylene glycol and naphthalene dicarboxylic acids

Polyethylene terephthalate (PET): A type of thermoplastic polymer with an ester linkage in its backbone and composed from ethylene glycol and terephthalic acid monomers

Polylactic acid: A type of biodegradable polyester derived from corn starch

Polymer: A large, long molecular chain

Polymer fines: Thin slivers of polymeric pellets known to have static attraction to the parts of an injection molding machine resulting in sliver decomposition and part discoloration

Polymer grade: A specific proprietary formulation provided by a manufacturer. It includes the polymer chains and additives.

Polymer laminate: A combination of polymeric layers, metallic layers, or both adhered together to form a single material structure

Polymer processing: The study and methodology of converting a polymer from one shape into another

Polyolefin: A thermoplastic composed of olefin monomers (C_nH_{2n})

Potting: A processing technique used with elastomers and thermosets—an enclosure is used to give the polymer shape, and it is left in place after the cure reaction is complete

Quantum efficiency curve: A curve that depicts the ratio of the total amount of electrons generated from the photovoltaic cell divided by the number of incident photons for each incident wavelength of light

Rheology: The study of material flow

Secant modulus: The slope, typically reported in pascal, of a line connecting the origin and a point on the stress-strain curve

Semiconductor: A chemistry that can conduct electrons under certain circumstances; the chemistry is typically doped with impurities to form electron acceptors (p-type) and electron donors (n-type)

Short circuit current: The current produced when there is zero voltage

Silicone: An elastomeric polymer composed of silicon and oxygen atoms in the backbone of the chain

Single crystalline silicon photovoltaic cell: The most common type of crystalline silicon photovoltaic cell composed of a single silicon crystal characterized by one long-range, uniform lattice structure with no grain boundaries; typical $E_g = 1.1$ eV

Single junction cell: A single cell composed of one semiconductor chemistry

Solar reflectance: The percentage of sunlight reflected from a surface, also known as albedo and measured in a scale from 0 to 1, with 1 indicating a 100% reflectance

Strain to break: The strain when the specimen macroscopically breaks

Superstrate: A material, typically glass, that is the top surface of the photovoltaic module

Tensile: A mode of deformation commonly used in bulk polymer mechanical testing described as pulling the same substrate simultaneously in opposite directions

Thermoplastic: A classification of polymers known for their high modulus and irreversible yield during deformation, also known as plastics

Thermoplastic elastomer: A classification of polymers which combines the rubbery properties of an elastomer and the thermoplastic's ability to flow at elevated temperature

Thin-film cells: A classification of photovoltaic cells denoted by the thin deposition (between a few nanometers and a few microns) of semiconductor material required to generate electricity from sunlight

Ultraviolet (UV) degradation: A chemical mechanism stimulated by ultraviolet (UV) radiation that causes a change in the polymer's chemical composition and physical properties

Ultraviolet-Visible (UV-Vis) spectrum: The transmission or absorbance of electromagnetic radiation by a substrate over the visible and ultraviolet wavelengths of the electromagnetic spectrum

Viscosity: The measure of a polymer's resistance to flow at a specified temperature and pressure while under shear or tensile stress

Index

Milton Keynes UK
Ingram Content Group UK Ltd.
UKHW031147141024
449569UK00024B/989